TABVLA IMPERII RO
CONDATE-GLEVVM-LONDINIVM-LVTETIA

UNION ACADÉMIQUE INTERNATIONALE

TABVLA IMPERII ROMANI CONDATE-GLEVVM-LONDINIVM-LVTETIA

BASED ON GSGS 4646 AT 1:1,000,000
(COVERING SHEET M.30 AND PART OF SHEET M.31
OF THE INTERNATIONAL 1:1,000,000 MAP OF THE WORLD)

LONDON
PUBLISHED FOR THE BRITISH ACADEMY
BY THE OXFORD UNIVERSITY PRESS

Oxford University Press, Walton Street, Oxford OX2 6DP
LONDON GLASGOW NEW YORK TORONTO
DELHI BOMBAY CALCUTTA MADRAS KARACHI
KUALA LUMPUR SINGAPORE HONG KONG TOKYO
NAIROBI DAR ES SALAAM CAPE TOWN
MELBOURNE AUCKLAND
AND ASSOCIATE COMPANIES IN
BEIRUT BERLIN IBADAN MEXICO CITY

Published in the United States by Oxford University Press, New York

ISBN 0 19 726020 9

© The British Academy 1983

Cover illustration *The front elevation of the Temple of Sulis Minerva*—Reproduced from *Roman Bath* (1969) by B. W. Cunliffe with kind permission of the Society of Antiquaries, London.

BASE MAP PRINTED BY
JOHN BARTHOLOMEW AND SON LIMITED
GAZETTEER PRINTED BY
LAMPORT GILBERT PRINTERS LIMITED

CONTENTS

Foreword — vii

Abbreviations and Bibliography — ix

Index — 17

Map of area around Amiens — 109

Plans of Caerleon, Caerwent, Cirencester, Colchester,
London, Paris, Silchester and Verulamium — *at end*

CONTRIBUTORS

R. AGACHE
D. BERTIN
R. BORIUS
R. CHEVALLIER
P. CRUMMY
J. DECAENS
R. W. FEACHEM
M. FLEURY
S. S. FRERE
M. G. FULFORD
R. A. GARDINER†

L. HARMAND
D. JALMAIN
P. LEMAN
B. LIGER
P. R. V. MARSDEN
P. MITARD
A. L. F. RIVET
M. SANQUER
N. H. H. SITWELL
J. B. WARD-PERKINS†

FOREWORD

As was explained in the Foreword to the provisional edition of Sheet M.31, published in 1975, this production represents a new development of the *Tabula Imperii Romani*, designed to make better use of the base-maps which are now available. While the scale of 1:1,000,000 has been retained, the sheet's coverage has been extended to the west to include the Isles of Scilly and to the east to include the western half of what was formerly Sheet M.31. The southern boundary remains the same, at 48°N, so as not to complicate the planning of the L series of maps further south, and it is hoped that at some future date the eastern half of Sheet M.31 will be combined with a new edition of Sheet M.32 (*Mogontiacum*), to replace the now outdated 1940 issue of the latter. A map of the rest of Britain already exists in the *Ordnance Survey Map of Roman Britain*, although this lacks a full bibliography and the fourth edition of it, published in 1978, is now at a scale of 1:625,000.

The design and production of the present map owe most to the late J. B. Ward-Perkins, whose devoted work both as President of the Central Committee of the *Tabula Imperii Romani* and as Chairman of its British Committee will long be remembered. As with the provisional edition of Sheet M.31, the collection of the French material has been in the hands of M. Raymond Chevallier, with assistance in this case from Mlle. D. Bertin and MM. R. Agache, R. Borius, J. Decaens, M. Fleury, L. Harmand, D. Jalmain, P. Leman, B. Liger, P. Mitard and M. Sanquer. The collection of the British material, and its marrying with the French, has been the work first of Mr. R. W. Feachem (then of the Ordnance Survey) and subsequently of Mr. N. H. H. Sitwell, with assistance on points of detail from Mr. P. Crummy, Dr. M. G. Fulford and Mr. P. R. V. Marsden. For cartographic advice in the earlier stages we owed much to the late Brig. R. A. Gardiner (then Keeper of the Map Room of the Royal Geographical Society). The main map has been drawn by Mr. D. Cox and the town-plans by Mr. P. and Mrs. V. Winchester. We are grateful to M. J. Picard, the publisher of the provisional Sheet M.31, for permission to reprint entries from the Index to that map.

While an attempt has been made to bring up to date the material transferred from the provisional sheet, we have here omitted two categories which were there included. These are, first, native hill-forts, because in some areas (most notably in western Britain) their inclusion would overcrowd the map and obscure the subsequent Roman development; and second, isolated finds, since their discovery is somewhat random and a distribution of coin-hoards which excludes those found on occupied sites can be misleading. This is not, of course, to say that the distribution of monuments and sites which are included can be taken as final. A great deal has been discovered even in the past eight years and both in Britain and in Gaul there are certainly a number of military camps and forts relating to the initial Roman occupation still to be found. As for villas, the remarkable density revealed in the

Somme basin by the brilliant and intensive aerial investigations of M. Roger Agache must lead one to wonder what a similar operation could produce elsewhere. A further limitation of a map such as this arises from the fact that it covers a period of about half a millennium and it is difficult to distinguish the different stages of development and decline.

In Gaul the boundaries of provinces can be shown, not only those of the earlier Gallia Belgica and Gallia Lugdunensis but also those of the later subdivisions, first (as in the Verona List) between Belgica Secunda, Lugdunensis Prima and Lugdunensis Secunda, and later (as in the *Notitia Galliarum*) between a slightly-reduced Belgica Secunda, a much-reduced Lugdunensis Secunda, Lugdunensis Tertia and Lugdunensis Senonia. That this is possible is due mainly to the evidence of the *Notitia Galliarum*, coupled with the fact that to a large extent the bishoprics of medieval France reflected the tribal *civitates* of an earlier age. No such definition can be attempted in Britain: here, the most that can be said is that the whole of the area shown on this map fell within the Severan province of Britannia Superior and that Cirencester was almost certainly the capital of the later province of Britannia Prima and London that of Maxima Caesariensis. The boundary between the two remains a matter for speculation.

A. L. F. RIVET
Chairman
British T.I.R. Committee

London
January, 1983

ABBREVIATIONS AND BIBLIOGRAPHY

I. Ancient Sources and Commentaries

Amm. Marc.	Ammianus Marcellinus, *Res Gestae*.
Aur. Vic.	Aurelius Victor.
Aus.	Ausonius.
Beda, *Ecc. Hist.*	Bede, *Historia Ecclesiastica Gentis Anglorum*.
Caes. *BG*	Caesar, *De Bello Gallico*.
Cass. Dio	Cassius Dio, *Historiae Romanae*.
Cassiod.	Cassiodorus.
Chron. Fredeg.	*Chronica Fredegarii*.
Cic. *ad F.*	Cicero, *Epistulae ad Familiares*.
CIIC	Macalister, R. A. S., *Corpus Inscriptionum Insularum Celticarum*, Dublin, 1945–49.
CIL	*Corpus Inscriptionum Latinarum*.
Cod. Theod.	Codex Theodosianus.
Conc. Arel. a. 314	
Conc. Agrip. a. 346	
Conc. Paris. a. 360	
Conc. Aurel. a. 511	Councils of Arles, AD 314; Cologne, 346; Paris, 360; Orléans, 511. *Cf Concilia Galliae*, Turnhout, 1963.
Diod. Sic.	Diodorus Siculus.
Edict. Diocl.	*Edictum Diocletiani de Pretiis*.
Eutrop.	Eutropius.
Flor.	Florus, *Epitomae de Tito Livio*.
GGM	Müller, K. (ed.), *Geographi Graeci Minores*, Paris, 1861; reprinted Hildesheim, 1965.
Gildas	Gildas Sapiens, *De Excidio et Conquestu Britanniae*.
GLM	Riese, A., (ed.), *Geographi Latini Minores*, Berlin, 1878; reprinted Hildesheim, 1964.
Greg. Tur. *HF*	Gregory of Tours, *Historia Francorum*.
Head	Head, B. V., *Historia Nummorum*, Oxford, 1911.
Hier.	S. Hieronymus (St Jerome).
Hist. Brit.	*Historia Brittonum*.
Isidorus	Isidorus Hispalensis (Isidore of Seville).
It. A.	*Itinerarium Provinciarum Antonini Augusti* (the Antonine Itinerary), ed. O. Cuntz, Leipzig, 1929.
It. M.	*Itinerarium Maritimum.* (the Maritime Itinerary), ed. O. Cuntz.
Jordanes	Jordanes, *Getica*.
Jul. Hon.	Julius Honorius, *Cosmographia* (text, *cf. GLM*).
Julian.	Julianus (emperor, AD 360–63).
Juv.	Juvenal, *Satires*.

Liv.	Livy, *Ab Urbe Condita*.
Lucan.	Lucan, *Pharsalia*.
Mack	Mack, R. P., *The Coinage of Ancient Britain*, 2nd edition, London, 1964.
Marcian.	Marcianus Heracleensis, *Periplus Maris Exteri* (text, *cf GGM*).
Mart.	Martial.
Mela	Pomponius Mela, *De Chorographia*.
Not. Dig. Occ./Or.	*Notitia Dignitatum Occidentis/Orientis*.
Not. Gall.	*Notitia Galliarum*.
Not. Tiron.	*Notae Tironianae*.
Oros.	Orosius.
Paneg.	*Panegyrici Latini*, ed. R. A. B. Mynors, Oxford, 1964.
Paulinus Nol.	Paulinus of Nola.
Plin. *NH*	Pliny the Elder, *Naturalis Historia*.
Plut.	Plutarch.
PNRB	Rivet, A. L. F. and Smith, C., *The Place-Names of Roman Britain*, London, 1979.
Ptol.	Claudius Ptolemaeus, *Geographia*.
Rav. Cos.	*Ravennatis Anonymi Cosmographia*, ed. J. Schnetz, Leipzig, 1940.
RIB	Collingwood, R. G. and Wright, R. P., *The Roman Inscriptions of Britain I*, Oxford, 1965.
Seneca, *Apocol.*	Seneca, *Apocolocyntosis*.
SHA	*Scriptores Historiae Augustae*.
Sid. Apoll.	Sidonius Apollinaris.
Sil. Ital.	Silius Italicus.
Solinus	Solinus, *Collectanea Rerum Memorabilium*.
Soz.	Sozomenus.
Steph. Byz.	Stephanus of Byzantium (Étienne de Byzance).
Str.	Strabo, *Geographia*.
Suet.	Suetonius.
Sulp. Sev.	Sulpicius Severus, *Chronica*.
Tab. Peut.	*Tabula Peutingeriana* (the Peutinger Table).
Tac. *Agric., Ann., Hist.*	Tacitus, *Agricola, Annales, Historiae*.
Tert.	Tertullian.
Ven. Fort.	Venantius Fortunatus.
Vib. Seq.	Vibius Sequester.
Wuilleumier	Wuilleumier, P., Inscriptions latines des Trois Gaules, *Gallia Supp.* XVII, Paris, 1963.
Zos.	Zosimus.

II. Modern Sources

AC	*L'Antiquité classique*, Brussels.
ACN	*Annuaire des Côtes du Nord*, Saint-Brieuc.
Agache, fig.	Agache, R., Vues aériennes de la Somme et recherche du passé, *BSPN* V (1962), fig. 1–93; Archéologie aérienne de la Somme, recherches nouvelles, 1963–4, *ibid.* VI (1964), fig. 94–271; Détection aérienne des vestiges protohistoriques, gallo-romaines et médiévaux dans le bassin de la Somme et de ses abords, *ibid.* VII (1970), fig. 272–658.
Agache, 1972	*Idem*, Quelques *fana* repérés par avion dans le bassin de la Somme et en Artois, *BSAP*, 1972, 301–324.
Agache, 1978	*Idem, La Somme pré-romaine et romaine*, Amiens.
Agache-Bréart, 1975	Agache, R. and Bréart, B., *Atlas d'archéologie aérienne de Picardie*, Amiens.
Album Comm Ant SI	*Album de la Commission des antiquités de la Seine-Inférieure*, Rouen.
ANL	*The Archaeological News Letter*, London.

Ann Ass Norm	*Annuaire de l'Association normande.*
Ann Bret	*Annales de Bretagne*, Rennes.
Ann Manche	*Annuaire de la Manche.*
Ann Norm	*Annales de Normandie*, Caen.
Ann 5 Norm	*Annuaire des cinq départements de la Normandie*, Caen.
Ant	*Antiquity*, Cambridge.
Ant J	*The Antiquaries Journal*, London and Oxford.
Arch	*Archaeologia*, London.
Arch Camb	*Archaeologia Cambrensis*, Cardiff.
Arch Cant	*Archaeologia Cantiana*, Maidstone.
Arche	*Archéologia*, Paris.
Arch J	*The Archaeological Journal*, London.
Arm	*L'Armoricain.*
ASHAStM	*Annales de la Société d'histoire et d'archéologie de Saint-Malo.*
ASPM	cf *BSPM.*
Asselin, 1829	Asselin, A., *Notice sur la découverte des restes d'une habitation romaine dans la mielle de Cherbourg et sur d'autres antiquités trouvées de nos jours dans les arrondissements de Valognes et Cherbourg*, Cherbourg.
Athenaeum	*The Athenaeum*, London.
BACTH	*Bulletin archéologique et historique du Comité des travaux historiques*, Paris.
BAR	*British Archaeological Reports.*
Baucat	Baucat, P., Étude sur les voies romaines du département d'Ille-et-Vilaine, *BSAIV* LIV (1927), 3–82.
BAVF	*Bulletin archéologique du Vexin français.*
BBCS	*Bulletin of the Board of Celtic Studies*, Cardiff.
BCASM, BCASI	*Bulletin de la Commission des antiquités de la Seine-Maritime (-Inférieure)*, Rouen.
BCDMHPC, MCDMHPC	*Bulletin (Mémoires) de la Commission départementale des monuments historiques du Pas-de-Calais*, Arras.
BCHN	*Bulletin de la Commission historique du département du Nord*, Lille.
Berks AJ	*Berkshire Archaeological Journal*, Reading.
Besnier	Besnier, M., *Lexique de géographie ancienne*, Paris, 1914.
BGASM	*Bulletin du groupement archéologique de Seine-et-Marne*, Melun.
Blanchet, *1900*	Blanchet, A., *Les trésors de monnaies romaines et les invasions germaniques en Gaule*, Paris.
Blanchet, *Enceintes*	Idem, *Les enceintes romaines de la Gaule*, Paris, 1907.
Blanchet, 1908	Idem., *Recherches sur les aqueducs et cloaques de la Gaule romaine*, Paris.
BMQ	*British Museum Quarterly*, London.
Bouhier, 1962	Bouhier, C., *Inventaire des découvertes archéologiques du département de la Manche (périodes gallo-romaine et mérovingienne)*, thesis, University of Caen.
Bouton	Bouton, A., *Le Maine, histoire économique et sociale*, Le Mans, 1962.
Branigan-Fowler, 1976	Branigan, K., and Fowler, P.J. (ed.), *The Roman West Country*, Newton Abbot.
Britannia	*Britannia, a Journal of Romano-British and Kindred Studies*, London.
BSAAM	*Bulletin de la Société académique des antiquaires de la Morinie*, Saint-Omer.
BSAB	*Bulletin de la Société académique de l'arrondissement de Boulogne-sur-Mer.*
BSAEL	*Bulletin de la Société archéologique d'Eure-et-Loir*, Chartres.
BSAF	*Bulletin de la Société archéologique du Finistère*, Quimper.
BSAIC	*Bulletin de la Société artistique et industrielle de Cherbourg.*
BSAIV	*Bulletin de la Société archéologique du département d'Ille-et-Vilaine*, Rennes.
BSAL	*Bulletin de la Société archéologique de la Loire-Inférieure*, Nantes.
BSAN, MSAN	*Bulletin (Mémoires) de la Société des Antiquaires de Normandie*, Caen.
BSAP, MSAP	*Bulletin (Mémoires) de la Société des Antiquaires de Picardie*, Amiens.
BSCN, MSCN	*Bulletin (Mémoires) de la Société d'émulation des Côtes du Nord*, Saint-Brieuc.

BSÉA, MSÉA	Bulletin (Mémoires) de la Société d'émulation historique et littéraire d'Abbeville.
BSFFA	Bulletin de la Société française des fouilles archéologiques, Paris.
BSFN	Bulletin de la Société française de numismatique, Paris.
BSHADM	Bulletin de la Société d'histoire et d'art du diocèse de Meaux.
BSHAO	Bulletin de la Société historique et archéologique de l'Orne, Alençon.
BSNAF, MSNAF	Bulletin (Mémoires) de la Société nationale des Antiquaires de la France, Paris.
BSNÉP	Bulletin de la Société normande d'études préhistoriques, Rouen.
BSPF	Bulletin de la Société préhistorique française, Paris.
BSPM	Bulletin de la Société polymathique du Morbihan, Vannes.
BSPN	Bulletin se la Société préhistorique du Nord, Amiens.
Bull Arch	Bulletin archéologique, Paris.
Bull Arch Ass Bret	Bulletin archéologique de l'Association bretonne, Rennes.
Bull Com Ant P-de-C	Bulletin de la Commission des antiquités départementales du Pas-de-Calais, Arras.
Bull Inst Arch Lond	Bulletin of the Institute of Archaeology (London).
Bull Mon	Bulletin monumental, Paris.
Bull Soc agr Sarthe	Bulletin de la Société d'agriculture, sciences et arts de la Sarthe, Le Mans.
Bull Soc Arch Orl	Bulletin de la Société archéologique et historique de l'Orléanais, Orléans.
Bull Soc Brest	Bulletin de la Société academique de Brest.
Bull Soc Clermont	Bulletin (Mémoires, Comptes rendus) de la Société archéologique et historique de Clermont-en-Beauvaisis.
Bull Soc Creil	Bulletin de la Société archéologique, historique et géographique de Creil et sa région.
Bull Soc Émul SI	Bulletin de la Société d'émulation de la Seine-Inférieure, Rouen.
Bull Soc Géo Lille	Bulletin de la Société géographique de Lille.
Bull Soc sc Yonne	Bulletin de la Société des sciences historiques et naturelles de l'Yonne, Auxerre.
Bull Som Arch	Bulletin of Somerset Archaeology, Taunton.
CAGQB	Colchester Archaeological Group Quarterly Bulletin, Colchester.
Cambry, 1803	Cambry, J. de, Description du département de l'Oise, Paris.
C Arch	Current Archaeology, London.
Carm Ant	The Carmarthen Antiquary, Carmarthen.
Cayot-Delandre	Cayot-Delandre, F.M., Le Morbihan, son histoire et ses monuments, Vannes and Paris, 1847.
CBA	Council for British Archaeology.
Celticum	Celticum, supplement to Ogam, Rennes.
Cochet, Rép.	Cochet, Abbé, Répertoire archéologique du département de la Seine-Inférieure, Paris, 1871.
Cochet, S.I.	Idem, La Seine-Inférieure historique et archéologique, 2nd edition, Paris, 1866.
Colbert de Beaulieu	Colbert de Beaulieu, J.-B., Traité de numismatique celtique, Paris, 1973.
Coll Bret	Le Collectionneur breton.
Collingwood-Richmond, 1969	Collingwood, R.G., and Richmond, I., The Archaeology of Roman Britain, London.
Comm S–O	Bulletin de la Commission des antiquités et arts du département de Seine-et-Oise, Versailles.
Corn Arch	Cornish Archaeology, Truro.
Coutil	Coutil, L., Archéologie gauloise, romaine, franque et carolingienne du département de l'Eure, Paris, 1895–1925.
CRAI, MAI	Comptes rendus (Mémoires) de l'Académie des inscriptions et belles-lettres, Paris.
CRCAF	Comptes rendus du Congrès archéologique de la France.
Cunat, 1850	Cunat, C., Deuxième rapport au Préfet des Côtes du Nord, Saint-Malo.
Cunliffe, 1968	Cunliffe, B., (ed.), Fifth Report on the Excavations of the Roman Fort at Richborough, Kent, London and Oxford.

Cunliffe, 1973	Idem, *The Regni*, London.
CVP	*Dossiers et procès-verbaux de l'Inspection des fouilles archéologiques de la Commission du vieux Paris.*
DAG	*Dictionnaire archéologique de la Gaule*, Paris, 1870–1919.
de Caumont	de Caumont, A., *Cours d'antiquités monumentales*, Paris, 1831–8; *Statistique monumentale du Calvados*, Caen, 1846–67; *Abécédaire ou Rudiment d'archéologie, Ère gallo-romaine*, 2nd edition, Caen, 1870.
de Courcy	de Courcy, P., *Notice historique sur la ville de Landerneau*, 1906.
de Gerville	de Gerville, C., *Études géographiques et historiques sur le département de la Manche*, Cherbourg, 1854.
Deglatigny, 1925/7	Deglatigny, L., *Documents et notes archéologiques*, Rouen, vol. I, 1925; vol; II, 1927.
Deglatigny, 1931	Idem, *Inventaire archéologique de la Seine-Inférieure, Période gallo-romaine*, Évreux.
de la Barre de Nanteuil	de la Barre de Namteuil, A, *Guide du Congrès de Brest-Vannes de la Société française d'archéologie*, 1914.
Delmaire, 1976	Delmaire, R., *Étude archéologique de la partie orientale de la cité des Morins*, Arras.
Dérolez	Dérolez, A., La cité des Atrébates à l'époque gallo-romaine, documents et problèmes, *RN* XL (1958), 505-33.
Desjardins, *Géo Gaule*	Desjardins, E., *Géographie historique et administrative de la Gaule romaine*, Paris, 1876–93.
Desjardins, *Tab Peut*	Idem, *La Table de Peutinger*, Paris, 1869–74.
Doranlo	Doranlo, R., *Épigraphie Lexovienne*, 1931.
du Chatellier	du Chatellier, P., *Les époques préhistoriques et gauloises dans le Finistère*, Rennes and Quimper, 1907.
Dudley-Webster, 1965	Dudley, D. R. and Webster, G., *The Roman Conquest of Britain*, London.
du Fretay, 1898	du Fretay, H., *Histoire du Finistère de la formation quaternaire à la fin de l'ère romaine*, Quimper.
Dunnett, 1975	Dunnett, R., *The Trinovantes*, London.
Durvin	Durvin, P., *La région de Creil à l'époque gallo-romaine*, Beauvais.
Duval, 1961	Duval, P.-M., *Paris antique des origines au IIIe siècle*, Paris.
EHR	*English Historical Review*, London.
Espérandieu	Espérandieu, E., *Recueil général des bas-reliefs, statues et bustes de la Gaule romaine*, Paris.
Et G et H Manche	de Gerville, C., *Études géographiques et historiques sur le département de la Manche*, Cherbourg, 1854.
Fossier	Fossier, R., *La terre et les hommes en Picardie jusqu'à la fin du XIIIe siècle*, Paris and Louvain, 1968.
Fouquet	Fouquet, A.?, *Des monuments celtiques et des ruines romaines dans le Morbihan*, Vannes, 1853.
Fox, 1923	Fox, C., *The Archaeology of the Cambridge Region*, Cambridge.
Frere, 1978	Frere, S. S., *Britannia*, revised edition, London.
Gallia	*Gallia; Fouilles et monuments archéologiques en France métropolitaine*, Paris.
Geog J	*The Geographical Journal*, London.
Graves, 1856	Graves, L., *Notice archéologique sur le département de l'Oise*.
Grenier, *Manuel*	Grenier, A., *Manuel d'archéologie gallo-romaine*, Paris, 1931–60.
Guennou	Guennou, G., *La cité des Coriosolites*, Rennes, 1981.
Habasque, 1834/1836	Habasque, X., *Notions historiques, géographiques, etc., sur le littoral des Côtes du Nord*, Saint-Brieuc.
Hawkes-Hull, 1947	Hawkes, C. F. C. and Hull, M. R., *Camulodunum: First Report on the Excavations at Colchester, 1930–1939*, London and Oxford.
Helinium	*Helinium*, Wetteren, Belgium.
Herts Arch	*Hertfordshire Archaeology*, St Albans.
Hull, 1958	Hull, M. R., *Roman Colchester*, Oxford.
Hull, 1963	Idem, *The Roman Potters' Kilns of Colchester*, Oxford.

Jalmain	Jalmain, D., *Archéologie aérienne en Île-de-France*, Paris, 1970.
Jarrett-Dobson, 1965	Jarrett, M. G. and Dobson, B. (ed.), *Britain and Rome*, Kendal.
JBAA	*Journal of the British Archaeological Association*, London.
Jelski	Jelski, G., Six ans de recherches archéologiques en Artois, *BCDMHPC* IX.2 (1972–3), 131 seq.
Jessup, 1970	Jessup, R., *South-East England*, London.
Johnson, 1979	Johnson, S., *The Roman Forts of the Saxon Shore*, 2nd edition, London.
Johnston, 1977	Johnston, D. E. (ed.), *The Saxon Shore*, CBA Research Report 18, London.
JRIC	*Journal of the Royal Institution of Cornwall*, Truro.
JRS	*Journal of Roman Studies*, London.
JS	*Journal des Savants*, Paris.
Jullian, *HG*	Jullian, C., *Histoire de la Gaule*, Paris, 1909–26.
KAR	*Kent Archaeological Review*, Dover.
LA	*The London Archaeologist*, London.
Lambrechts, 1942	Lambrechts, P., *Contributions à l'étude des divinités celtiques*, Bruges.
Langouet, 1973	Langouet, L., *Alet, ville ancienne*, thesis, University of Rennes, unpublished.
Lantier	Lantier, R., *Recueil général . . .*, continuation of Espérandieu, qv.
Latomus	*Latomus, revue d'études latines*, Brussels.
Ledru, 1911	Ledru, A., *Répertoire des monuments et objets anciens de la Sarthe*, Le Mans.
Leduque, *Ambianie*	Leduque, A., *Esquisse de topographie historique sur l'Ambianie*, Amiens, 1972.
Leduque, *Atrébatie*	Idem, *Recherches topo-historiques sur l'Atrébatie*, Lille, 1966.
Leduque, *Boulonnais*	Idem, *Étude sur l'ancien réseau routier du Boulonnais*, Lille, 1957.
Leduque, *Morinie*	Idem, *Essai de topographie historique sur la Morinie*, Lille, 1968.
Le Mené, 1891	Le Mené, *Histoire archéologique, féodale et religieuse des paroisses du diocèse du Morbihan*, Vannes.
Le Prévost, 1862	Le Prévost, *Mémoires et notes pour servir à l'histoire du département de l'Eure*.
Leroux	Leroux, A., Note sur le chatellier industriel à Langonnet, *BSAL* XLV (1904), 13–47.
Lewis, 1966	Lewis, M. J. T., *Temples in Roman Britain*, Cambridge.
McWhirr, 1981	McWhirr, A., *Roman Gloucestershire*, Gloucester.
MAI	cf *CRAI*.
Mathière, 1925	Mathière, J., *La civitas des Aulerci Eburovices à l'époque gallo-romaine*, Évreux.
MÉFR	*Mélanges de l'École française de Rome*.
Mém Soc Clermont	cf *Bull Soc Clermont*.
Mém Soc Senlis	*Mémoires de la Société d'histoire et d'archéologie de Senlis*.
Mém Soc Val	*Mémoires de la Société archéologique, artistique, littéraire et scientifique de l'arrondissement de Valognes*.
Miller, 1916	Miller, K., *Itineraria Romana*, Stuttgart.
Mon Ant	*The Monmouthshire Antiquary*, Monmouth.
Monuments Piot	Fondation Eugène Piot, *Monuments et Mémoires*, Académie des inscriptions et belles-lettres, Paris.
Morgannwg	*Morgannwg*, Cardiff.
MPBWArchExc	Ministry of Public Building and Works, *Archaeological Excavations Annual Reports*.
MSAN, MSAP, etc.	cf *BSAN, BSAP*, etc.
MSHAB	*Mémoires de la Société d'histoire et d'archéologie de Bretagne*, Rennes.
Nash-Williams, 1969	Nash-Williams, V. E., *The Roman Frontier in Wales*, 2nd edition (ed. M. G. Jarrett), Cardiff.
Nenquin	Nenquin, J., *Salt – a study in economic prehistory*, Bruges, 1961.
Nouel	Nouel, A., *Les origines gallo-romaines de la région de la Loire moyenne et de ses abords*, Gien, 1968.
Ogam	*Ogam*, Rennes.
Orain, 1882	*Géographie pittoresque du département d'Ille-et-Vilaine*, Rennes.

Oxon	Oxoniensia, Oxford.
Pape, 1969	Pape, L., L'Armorique gallo-romaine, in J. Delumeau (ed.), *Histoire de la Bretagne,* Toulouse, 1969, 90–115.
Paris et Île-de-France	Mémoires publiés par la Fédération des Sociétés historiques et archéologiques de Paris et de l'Île-de-France, Paris.
Pauthiel, 1927	Pauthiel, E., *Notions d'histoire et d'archéologie pour la région de Fougères,* Rennes.
PBN	Le Pays Bas-Normand, Flers, Orne.
PBNHAFC	Proceedings of the Bath Natural History and Archaeological Field Club, Bath.
PDAS	Proceedings of the Devon Archaeological Society, Exeter.
PDNHAS, PDNHAFC	Proceedings of the Dorset Natural History and Archaeological Society (Field Club), Dorchester, Dorset.
Pesche, *Dictionnaire*	Pesche, J.-R., *Dictionnaire topographique de la Sarthe,* Le Mans and Paris, 1829–42.
PHFC	Proceedings of the Hampshire Field Club, Winchester.
PIOWNHAS	Proceedings of the Isle of Wight Natural History and Archaeological Society, Newport, Isle of Wight.
Ponchon, 1913	Ponchon, A., *Les villas gallo-romaines en Picardie,* Amiens.
Prarond	Prarond, E., *Histoire de cinq villes et de 300 villages,* Abbeville, 1861–8.
PSAL	Proceedings of the Society of Antiquaries of London, London.
PSANHS	Proceedings of the Somerset Archaeological and Natural History Society, Taunton.
PUBSS	Proceedings of the University of Bristol Spelaeological Society, Bristol.
P-W	Pauly, A.F. von, Wissowa, G., *et al., Real-Encyclopädie der Classischen Altertumswissenschaft,* Stuttgart.
RA	Revue archéologique, Paris.
RAE	Revue archéologique de l'Est et du Centre-Est, Dijon.
Rainey, 1973	Rainey, A., *Mosaics in Roman Britain: a gazetteer,* Newton Abbot.
RBPh	Revue belge de philologie.
RCAHM (Wales)	Royal Commission on Ancient and Historical Monuments (Wales).
RCHM	Royal Commission on Historical Monuments. (England).
RÉA	Revue des études anciennes, Bordeaux.
Records of Bucks	Records of Buckinghamshire, Aylesbury.
RÉL	Revue des études latines, Paris.
Renault, 1880	Renault, C., *Inventaire des découvertes préhistoriques et gallo-romaines faites dans les environs de Cherbourg,* Cherbourg.
Rev Arm	Revue de l'Armorique.
Rev Av	Revue de l'Avranchin, Avranches.
Rev de Norm	Revue de Normandie.
Rev Manche	Revue du département de la Manche, Saint-Lô.
Rev Num	Revue numismatique, Paris.
Rev Prov Ouest	Revue des provinces de l'Ouest.
Rev Soc Arch Orl	of Bull Soc Arch Orl.
RH	Revue historique.
RHAM	Revue historique et archéologique du Maine, Le Mans.
Ringot	Ringot, R., Sites et trouvailles d'époque celtique et gallo-romaine en Morinie septentrionale (région Calais-Saint-Omer), *Celticum* XV (1966), 151 *seq.*
RIO	Revue internationale d'onomastique, Paris.
Rivet, 1964	Rivet, A.L.F., *Town and Country in Roman Britain, revised edn.,* London.
Rivet, 1969	*Idem* (ed), *The Roman Villa in Britain,* London.
RN	Revue du Nord, Lille.
Roblin, 1971	Roblin, M., *Le terroir de Paris aux époques gallo-romaine et franque,* 2nd edition, Paris.
Roblin, 1978	*Idem, Le terroir de l'Oise aux époques gallo-romaine et franque,* Paris.

Rodwell-Rowley, 1975	Rodwell, W., and Rowley, T. (ed.), *The 'Small Towns' of Roman Britain*, BAR 15, Oxford.
Ross, 1967	Ross, A., *Pagan Celtic Britain*, London.
RSSHN	*Revue des Sociétés savantes de Haute-Normandie*, Rouen.
Salway, 1981	Salway, P., *Roman Britain*, Oxford.
SASN	*Sussex Archaeological Society Newsletter*, Lewes.
SDNQ	*Somerset and Dorset Notes and Queries*, Sherborne.
Sedgley, 1975	Sedgley, J.P., *The Roman Milestones of Britain*, Oxford.
Septentrion	*Septentrion*, Calais.
Straker, 1931	Straker, E., *Wealden Iron*, London.
SxAC	*Sussex Archaeological Collections*, Lewes.
SxNQ	*Sussex Notes and Queries*, Lewes.
SyAC	*Surrey Archaeological Collections*, Guildford.
TBGAS	*Transactions of the Bristol and Gloucestershire Archaeological Society*, Gloucester.
TDA	*Transactions of the Devon Association*, Exeter.
TEAS	*Transactions of the Essex Archaeological Society*, Colchester.
TEHAS	*Transactions of the East Hertfordshire Archaeological Society*, Hertford.
TLMAS	*Transactions of the London and Middlesex Archaeological Society*, London.
TNFC	*Transactions of the Newbury District Field Club*, Newbury.
Todd, 1978	Todd, M. (ed.), *Studies in the Romano-British Villa*, Leicester.
Toussaint, *Seine*	Toussaint, M., *Répertoire archéologique du département de la Seine (moins Paris)*, Paris, 1953.
Toussaint, *S-O*	Idem, *Répertoire archéologique du département de Seine-et-Oise*, Paris, 1951.
Trevedy	Trevedy, J., *La pierre sculptée du Rillon, commune de Saint-Brandan, près de Quintin: un dieu au maillet*, 1892-95.
TStAHAAS	*Transactions of the St Albans and Hertfordshire Architectural and Archaeological Society*, St Albans.
TWNFC	*Transactions of the Woolhope Naturalists' Field Club*, Hereford.
Vasselle	Vasselle, F., Inventaire des constructions romaines rurales du département de la Somme, *Celticum* XII (1964), 289-332.
VCH	*The Victoria History of the Counties of England*.
Vesly, 1909	Vesly, L. de, *Les fana ou petits temples gallo-romains de la région normande*, Rouen.
Voisin, 1862	Voisin, A., *Les Cénomans anciens et modernes*, Paris.
Wacher, 1966	Wacher, J.S. (ed.), *The Civitas Capitals of Roman Britain*, Leicester.
Wacher, 1975	Idem, *The Towns of Roman Britain*, London.
WANHM	*The Wiltshire Archaeological and Natural History Magazine*, Devizes.
Webster, 1980	Webster, G., *The Roman Invasion of Britain*, London.
Webster, 1981	Idem, *Rome against Caratacus*, London.

INDEX

*For places marked with an asterisk see the map of the area around Amiens on page 109

ABBEVILLE Bel., *Ambiani* (Somme, F) 50°06′N, 1°50′E
'Camp de César', probable fort. Coin hoard, foundations, pottery. Musée Boucher de Perthes.
Louandre, C., *Histoire d'A.*, Abbeville, I, 1844; II, 1864: Agache, fig. 1, 31, 261, 280: Leduque, *Ambianie*, 83 sq: Agache-Bréart, 1975, 23: Agache, 1978, *passim*.

ABBOTTS ANN Brit., *Belgae* (Hampshire, GB) 51°10′N, 1°33′W
Villa; mosaics, painted plaster, pottery, tiles, coins late 3rd–early 4th c.
VCH Hampshire I (1900), 300: Rainey, 1973, 19.

ABERGAVENNY Brit., *Silures* (Gwent, GB) 51°49′N, 3°01′W
GOBANNIVM
It. A. 484.6 (*Gobannio*).
Rav. Cos. 106.25 (*Bannio*).
PNRB 369.
Fort, established c AD 50, abandoned c AD 150. Walls, granaries, bath-building.
Nash-Williams, 1969, 45.

ABONA cf Sea Mills

ABONA FLVMEN cf river Avon

ABRINCAS cf Avranches

ABRINCATVI
Caes. *BG* III, 9 (*Ambiliati*); VII, 75 (*Ambibarii*).
Plin. *NH* IV, 107 (*Abrincatuos*).
Ptol. II,8,8 ('Αβρινκάτουοι).
Not. Dig. Occ. V, 116 and 266; VII, 92 (*Abrincateni*); XXXVII, 11 and 22 (*Abrincatis*).
Not. Gall. II, 4 (*civitas Abrincatum*).
Oros. VI, 8 (*Ambivaritos*).
Conc. Aurel. a 511 (*episcopus ecclesiae Abrincatinae*).
Ven. Fort. *Vita S. Paterni* X, XII, XIII, XVI (*Abrincas*).
Greg. Tur. *Vita S. Martini* II, 36 (*Abrincatinae civitatis*); III, 19 (*Abrincatinus*): HF IX, 20 (*de Abrincatis*).
CIL XIII. 1, p 494.
Armorican tribe included in Lugdunensis (later Lugdunensis II), occupying the region of Avranchin with capital at INGENA, later known as ABRINCAS (Avranches, qv).

AD ANSAM cf Stratford St Mary

ADVRNI PORTVS cf Portus Ardaoni

AEPATIACI PORTVS cf Portus (A)epatiaci

AIGNEVILLE-HOCQUELUS Bel., *Ambiani* (Somme, F) 50°02′N, 1°36′E
Villas.
Vasselle, 294: Agache-Bréart, 1975, 23: Agache, 1978, 266.

AIRAINES Bel., *Ambiani* (Somme, F) 49°58′N, 1°57′E
At 'Les Coutumes' remains of a large villa: columns, capitals, hypocaust, kilns, tombs, tiles, pottery etc.
Marchand, Abbé A., *MSÉA* XXII (1908), 279–724; *Notes pour servir à l'histoire d'A*, 1909: Agache, figs 46, 85, 360, 546: Leduque, *Ambianie*, 160: Agache-Bréart, 1975, 23–24: Agache, 1978, 114, 146, 155, 173, 175, 179, 256, 262, 302, 345, 368, 419, 438.

AIRON-SAINT-WAAST Bel., *Morini* (Pas-de-Calais, F) 50°26′N, 1°40′E
At 'La Grosse Tombe' a cemetery.
BSAP VII (1882), 86, 415.

ALABVM cf ?Llandovery

ALAVNA cf Valognes

*****ALBERT** Bel., *Ambiani* (Somme, F) 50°00′N, 2°39′E
Villa and other buildings.
Agache-Bréart, 1975, 24: Agache, 1978, *passim*.

ALCHESTER Brit., *Catuvellauni* (Oxfordshire, GB) 51°53′N, 1°10′W
Town founded under Claudius (?); timber buildings reconstructed in stone c AD 100. Declined early 5th c. Walls, bath-building.
VCH Oxfordshire I (1939), 281–88: Rodwell-Rowley, 1975, 31, 118–123.

ALDENHAM Brit., *Catuvellauni* (Hertfordshire, GB) 51°42′N, 0°21′W
Villa; remains of walls, barn. Tile kiln built mid 2nd c, dismantled 3rd c. Detached bath building dated early 4th c.
JRS LIII (1963), 136; LVI (1966), 209.

ALDERNEY Lug.? Channel Islands (GB) ÎLE D'AURIGNY Les Îles Anglo-Normandes 49°42′N, 2°14′W
RIDVNA INSVLA

17

It. M. 509.2.
PNRB 181.
'The Nunnery', fort, probably part of the defensive system of the *Tractus Armoricanus*.
Johnston, 1977, 31–3: Johnson, 1979, 82, 151.

ALENÇON Lug., *Esuvii* or *Sagii* (Orne, F) 48°26'N, 0°05'E
Gallo-Roman foundations; *oppidum*. Musée de la Maison d'Oze.
BSHAO LXXVI (1958): *BSNAF*, 1966, 80.

ALETVM *cf* Saint-Malo

ALFOLDEAN Brit., *Regni* (West Sussex, GB) 51°05'N, 0°24'W
Minor settlement on Stane Street between London and Chichester, near a bridge over the river Arun (*Trisantona Fl.*). Rectangular earthwork enclosing c 1 ha. Occupied 1st to 4th c. Pottery, tiles, glass, tesserae, painted plaster, coins 1st to 4th c.
SxAC LXIV (1923), 81–104; LXV (1924), 112–57; LXXVI (1935), 183–92: *Arch J* CXVIII (1961), 163.

ALICE HOLT Brit., ?*Regni* or *Atrebates* (Hampshire, GB) 51°09'N, 0°51'W
Pottery, continuously active from Claudian period to 4th c.
SyAC LX (1963), 19–36: Wade, A. G., *Alice Holt, its History and Romano-British Potteries,* Guildford, 1949. M. A. B. Lyne and R. S. Jefferies: *The Alice Holt/Farnham Region Pottery Industry,* Oxford, 1979.

ALIZAY Lug., *Veliocasses* (Eure, F) 49°19'N, 1°11'E
Cemetery.
BCASI II (1870–72), 33–34.

ALLAINES Lug., *Carnutes* (Eure-et-Loir, F) 48°12'N, 1°50'E
At 'Le Marché de la Bouverie' the crossing of six roads, small settlement, foundations, coins, two villas.
Nouel, 9, 36: Jalmain, 95.

*ALLONVILLE Bel., *Ambiani* (Somme, F) 49°56'N, 2°21'E
Villa.
Agache, fig. 105, 275, 315, 651: *Gallia* XXIII (1965), 300: Vasselle, 296: Agache-Bréart, 1975, 24: Agache, 1978, 32, 102, 111, 112, 367, 456.

ALPHAMSTONE Brit., *Trinovantes* (Essex, GB) 51°59'N, 0°41'E
Tile kiln: roof tiles, hypocaust.
VCH Essex III (1963), 35–6.

ALRESFORD Brit., *Trinovantes* (Essex, GB) 51°49'N, 0°59'E
Corridor villa; frescoes, mosaics.
VCH Essex III (1963), 37: Dunnett, 1975, 96–8.

ALVERDISCOTT Brit., *Dumnonii* (Devon, GB) 50°59'N, 4°09'W
Temporary camp.
JRS LXVII (1977), 126.

AMAYÉ SUR ORNE Lug., *Viducasses* (Calvados, F) 49°05'N, 0°26'W

Villa on the road between Vieux and Lisieux.
de Caumont, 1846, 147: *BSAN* XXIX (1913–14), 221.

AMBIANI
Caes. *BG* II, 4, 15; VII, 75; VIII, 7.
Str. IV,3,5 (C. 194); IV,6,11 (C. 208) ('Αμβιανοί).
Liv. *Ep* CIV.
Plin. *NH* IV, 106.
Ptol. II,9,4 ('Αμβιανοί).
It. A. 362.4 (*Ambianis*).
Paneg. Const. Caes. XXI.
Conc. Agrip. a 346.
Amm. Marc. XV, 11, 10; XXVII, 8, 1.
Cod. Theod. VIII, 14, 1.
Not. Dig. Or. VI, 36 (*Ambianenses*); Occ. IX, 39 (*Ambianenses*); XLII, 67 (*Ambianos*).
Not. Gall. VI, 11 (*civitas Ambianensium*).
Oros. VI, 7 and 11.
Hier. *Ep ad Ageruchiam*; *Chron a Abraham 2383*.
Sulp. Sev. *Vita S. Martini* III, 1 (*Ambianensium civitas*).
Iul. Hon. *Cosmographia,* 19.
Not. Tiron. LXXXVII, 51.
Conc. Aurel. a 511 (*episcopus ecclesiae Ambianensis*).
Passio S. Iusti V, 374 (*Ambianis*); 376 (*Ambianensium civitatem*).
Acta Sanctorum Feb. 1, 198B; 11, 198E.
Ven. Fort. *Carmina* X, 6, 25; *Vita S. Martini* I, 56.
Greg. Tur. *Vita S. Martini* I, 17.
CIL VI, 11522 and 15493 (*Ambianae*); XIII.1, 607 (*civi Ambian[o]*); notes, p 549; XIII.2, 9032 (*C(ivitas) Amb(ianorum)*).
Belgic tribe, occupying Amiénois. *Oppida*: La Chaussée-Tirancourt, Chipilly, L'Étoile, Liercourt, Mareuil-Caubert. Capital: SAMAROBRIVA, later known as AMBIANIS (Amiens, *qv*). Other sites: Molliens-Vidame, Moliens-aux-Bois.
Leduque, *Ambianie, passim.*

AMFREVILLE-LA-MI-VOIE Lug., *Veliocasses* (Seine-Maritime, F) 49°24'N, 1°07'E
Gallo-Roman foundations, 'La Poterie': small finds, coins, 'Le Clos Madame'.
Cochet, S.I., 213: Vesly.

*AMIENS Bel., *Ambiani* (Somme, F) 49°54'N, 2°18'E
SAMAROBRIVA AMBIANORVM, later AMBIANIS.
Caes. *BG* V, 24, 47, 53 (*Samarobriva*).
Cic. *ad F* VII, 11, 2; 12, 1; 16, 3.
Ptol. II,9,4 (Σαμαροβρίουα/-βρίγα).
It. A. 362.4 (*Ambianis*), 379.9, 380.1 (*Samarobrivas*).
Tab. Peut. (*Sammarobriva*).
Not. Tiron. LXXXVII, 52 (*Samarobria*).
CIL XIII.1, pp 549–50 (notes); 3490 (altar, *Samarobrv[a]*); XIII.2, 9032 (milestone, *C(ivitas) Amb(ianorum) A S(amarobriva) L(euga) I*); 9158 (milestone of Tongres, *Samarabriva*).
Cf AMBIANI.
Important town, capital of the CIVITAS

AMBIANORVM. Roads to Boulogne, Arras, Soissons and Beauvais. Pre-Roman site at St-Acheul; Gallic settlement beside the bridge of Don (*Samarobriva* = 'Sommebridge'); *oppidum*, not well known, in the marsh or under the Cathedral. Gallo-Roman town occupied 40 ha in 1st c (cathedral sector), 120 ha in 2nd c, but with unoccupied spaces – gardens or a check in urbanisation? Regular street plan but poor construction. River harbour. Amphitheatre, 107×100 m; forum, baths, houses, sculptures, nine mosaics, many small finds including the 'Amiens *patera*' with information about Hadrian's Wall. Massive defensive wall (using the amphitheatre as a bastion) constructed after the first barbarian assault, AD 256; perimeter 1100 m, enclosed area 10 ha. Cavalry base of the late Empire (for *catafractarii*). Cemetery. Destroyed, AD 406.

Museums: Musée de Picardie, Bibliothèque municipale.

AC XIX (1950), 178–9; XXIV (1955), 148–9: Agache, fig. 16, 24–8, 60–61, 140, 253, 258, 285, 288, 324: Agache-Bréart, 1975, 25–6: *BACTH*, 1923, 3–17 (*cf Pro Alesia* IX–X (1923–4), 51–66, 138–70): Blanchet, *Enceintes*, 6, 120–22: *BSAP*, 1900, 557–8; 1901, 3, 338; 1916, 112–87; XXIX (1923), 511–9; XXXI (1925), 37–56, 448–68; XXXII (1928), 428–55; XXXIV (1933), 85–96; XXXV (1934), 111–9, 313–4, 529–50; XXXVI (1937), 114–8; XLIII (1951), 226–40, 352–74; XLIV (1953), 139; XLV (1953), 12–14; XLVI (1957), 16–46; XLVII (1960), 30–51, 205–25; XLVIII (1961), 165–83; L (1963–4), 65–8; LI (1965–6), 27–36: *BSNAF*, 1881, 246; 1952–3, 148–9: *Celticum* VI (1962), 323–42; IX (1963), 257–80; XV (1965), 185–205: Demailly, Catalogue des sigles de potiers gaulois et gallo-romains trouvés à Amiens depuis 25 ans, *MSAP* XXXVII (1914): Estienne, J., and Vasselle, F., *Le Bel Amiens*, A., 1967: *Gallia* VII (1949), 103–6; IX (1951), 72–8; XII (1954), 129–34; XIII (1955), 141–2; XIV (1956), 176; *Supp* X.1.1 (1957), nos 86–95; XVII (1959), 260–67; XX (1962), 79–101; XXV (1967), 202; XXIX (1971), 231; XXXI (1973), 344–6; XXXIII (1975), 310–3; XXXV (1977), 131–5; XXXVII (1979), 321–4: Gose, A., *Les enceintes successives d'Amiens*, A., 1853: Grenier, *Manuel* III.2 (1958), 697–8; IV.1 (1960), 328–31: Heurgon, J., La patère d'Amiens, *Monuments Piot*, 1952 (*cf JRS* XLI (1951), 22–4): *Hommages Grenier* III (1962), 1586–1600: *Latomus* VIII, 275; Leduque, *Ambianie*, 10, 13, 61–7: Massy, J.-L., *Amiens Gallo-Romain*, A., 1979: *Ogam* XXI (1969), 105: *RA*, 2nd series, XL (1880), 331; XLII (1881), 138: *RÉA* XXVIII (1925), 312–8: *RN* XXXVI (1954), 141–5, 147–9; XXXVII (1956), 320, 331–8; XL (1958), 467–82; XLII (1960), 337–52, 389–95; XLVIII (1966), 605–25; no 216 (Jan-March 1973), 29. *Cahiers archéologiques de Picardie*, No. 6 (1979), 109–29, 131–68.

ANCOURT Lug., *Caletes* (Seine-Maritime, F) 49°54′N, 1°11′E

Gallo-Roman foundations, burnt, near the church.
Cochet, *S.I.*, 253.

*****ANDECHY** Bel., *Ambiani* (Somme, F) 49°43′N, 2°43′E

Minor settlement on the road between Andechy and Le Quesnoy, cut by the Roye-Amiens road.
Agache, fig. 402, 403, 478: Leduque, *Ambianie*, 143: Vasselle, 296: Agache-Bréart, 1975, 26: Agache, 1978, 176, 258, 259, 353, 422.

ANDELYS (Les) Lug., *Veliocasses* (Eure, F) 49°15′N, 1°25′E

Foundations at 'Le Château'; at Noyers-sur-Andelys, 'La Fosse à dîme, Triage des Cateliers', a theatre – diameter 120 m, orchestra 60 m.
Gallia XVII (1959), 330.

ANDERITVM *cf* Andrésy, Pevensey

ANDIVM INSVLA *cf* ?Jersey

ANDRÉSY Lug., *Parisii* (Yvelines, F) 48°59′N, 2°04′E

?ANDERITVM.
Not. Dig. Occ. XLII, 22–3 (*praefectus classis Anderetianorum, Parisiis*).
Cf Pevensey.
?Port and naval station of the late Empire, on the river Seine. The identification is disputed.
Grenier, *Manuel* I (1931), 395 n 2: Toutain, J., Le problème d'Andrézy-sur-Seine, *BACTH*, 1941–2 (1946), 505–18: Duval, 1961, 229, 235 n 82: Roblin, 1971, 10 n 1, 47, 183: Johnson, 1979, 70.

ANGERVILLE Lug., *Carnutes* (Essonne, F) 48°19′N, 2°00′E

At 'Bois Chenu' 1500 m east of the town, a villa; bronze vessel and platters. At 'Poulainville' 800 m south of the town, a villa.
Air photograph, Jalmain: Duval, 1961, 90: Nouel, 33.

ANGLESQUEVILLE-L'ESNEVAL Lug., *Caletes* (Seine-Maritime, F) 49°38′N, 1°14′E

Cemetery.
Cochet, *S.I.*, 349.

ANGMERING Brit., *Regni* (West Sussex, GB) 50°50′N, 0°30′W

Villa; mosaics, large bath suite; built *c* AD 70.
VCH Sussex III (1935), 20: *SxAC* LXXX (1939), 88–92; LXXXVI (1947), 1–21: Cunliffe, 1973, 76–8.

ANNEVILLE/SEINE Lug., *Lexovii* (Seine-Maritime, F) 49°28′N, 0°53′E

Gallo-Roman foundations.
Cochet, *S.I.*, 175: *Gallia* XXVIII (1970), 275.

ANTIVESTAEVM sive **BELERIVM PROMONTORIVM** *cf* Land's End

ANVEVILLE Lug., *Caletes* (Seine-Maritime, F) 49°42′N, 0°44′E

'La Garenne', Gallo-Roman foundations, funerary urns, coin hoard Nerva to Alexander Severus.
Cochet, *S.I.*, 444: *BCASM* XV, 108.

APREMONT Bel., *Silvanectes* (Oise, F) 49°14′N, 2°31′E

'Braque', Forêt de la Haute-Pommeraye, Gallo-Roman foundations, wells.
Durvin.

AQVAE SEGESTAE *cf* Sceaux-du-Gâtinais

AQVAE SVLIS *cf* Bath

ARCEUIL *cf* Cachan

ARDAONI PORTVS *cf Portus Ardaoni*

ARDLEIGH Brit., *Trinovantes* (Essex, GB) 51°58′N, 0°59′E
Two pottery kilns, 2nd c.
VCH Essex III (1963), 38.

ARDRES Bel., *Morini* (Pas-de-Calais, F) 50°51′N, 1°59′E
At 'Les Terres Noires' a great assemblage of salt workings and settlement of the Gallo-Roman period; a road, buildings, cemetery, kilns, a dam. Earlier Gaulish settlement.
Cabal, P., Le site d'Ardres, *RN* LV (1973), 17–28: Leduque, *Morinie,* 79: Ringot, 176: Delmaire, 1976, *passim.*

AREGENVAE *cf* Vieux

ARGENTAN Lug., *Esuvii* or *Sagii* (Orne, F) 48°45′N, 0°01′W
'Faubourg de La Gaze', *vicus* or minor settlement; 3rd c coins.
MSAN IX (1835), 431–54, 462; XI (1838), 60.

ARGENTEUIL Lug., *Parisii* (Val d'Oise, F) 48°57′N, 2°15′E
ARGENTOGILVM (7th c).
Villa, aqueduct, cemetery.
Duval, 1961, 78, 231, 243, 284: Roblin, 1971, 25, 124, 158, 180, 204.

ARGOL Lug., *Osismii* (Finistère, F) 48°15′N, 4°19′W
'Treseulom', Gallo-Roman foundations, lime kiln.
BSAF, 1875, 122; 1972, 423.

ARGOUGES Lug., *Abrincatui* (Manche, F) 48°30′N, 1°24′W
Gallo-Roman foundations, coins, sherds, mosaics.
de Gerville, 1854, 81: *BSAIC* XXIV (1900), 152.

ARICONIVM *cf* Weston-under-Penyard

ARNE HEATH Brit., *Durotriges* (Dorset, GB) 50°41′N, 2°04′W
Salt working, 1st to 4th c.
RCHM Dorset II.3 (1970), 593.

ARNIÈRES-SUR-ITON Lug., *Aulerci Eburovices* (Eure, F) 48°59′N, 1°07′E
Theatre.
Grenier, *Manuel* III.2 (1958), 955.

AROSFA GAREG Brit., *Silures* (Dyfed, GB) 51°55′N, 3°44′W
Temporary camp, probably constructed during the campaign of Julius Frontinus, *c* AD 75; 18 ha.
BBCS XXI (1966), 174–78: Nash-Williams, 1969, 124–5; Frere, 1978, 121.

ARPAJON Lug., *Parisii* (Essonne, F) 48°35′N, 2°15′E

Formerly Châtres. Settlement of 1st c, *castrum* of the late Empire: cemetery.
Blanchet, *Enceintes,* 234: *Gallia* VII (1949–51), 111: Duval, 1961, 229, 261: Roblin, 1971, 8, 14, 99, 117, 301.

ARQUES Lug., *Caletes* (Seine-Maritime, F) 49°53′N, 1°08′E
'Archelles', a major settlement and cross-roads.
Cochet, *S.I.,* 248–51.

ARRAS Bel., *Atrebates* (Pas-de-Calais, F) 50°17′N, 2°48′E
NEMETACVM or NEMETOCENNA, later ATRABATIS.
Caes. *BG* VIII, 46 and 52 (*Nemetocenna*).
Ptol. II,9,4 (Μέτακον).
It. A. 377.8, 378.10, 379.2, 379.9 (*Nemetacum*).
Tab. Peut. (*Nemetaco*).
Not. Dig. Occ. XLII, 40 (*praefectus laetorum Batavorum Nemetacentium, Atrabatis*).
CIL XIII.1, p 558 (notes); XIII.2, 9158 (milestone of Tongres, *Nemetac*(*um*)); *cf* nos 603, 806 (*Nemetogena, Nemetocena,* women's names).
Cf ATREBATES.
Town, capital of the CIVITAS ATREBATVM. Roads to Amiens, Thérouanne, Tournai and Cambrai. Religious centre (the name N. means 'sacred grove') with commercial and political centre perhaps in the Quartier de Meaulens (*MEDIOLANVM; *cf* Évreux and Vieil-Évreux). Forum in the Place de la Préfecture?; temple under the cathedral?; large building of mid 1st c, burnt 172–4; houses, cellar of late 1st–2nd c, rooms, well of 3rd c, pottery, rubbish. *Castrum* of the late Empire; road, late tombs, shops, quarries. Four km distant, a military base, 'Mont de César'.
Museum of the Palais St-Vaast.
Bull Com Ant P-de-C VI (1885–8), 69: *BSNAF,* 1904, 403–9: Blanchet, *Enceintes,* 110–11, 255: *BCDMHPC* V (1922–30), 59; IX.1 (1971), 1–38, 60–72: Deneck, G., Les origines de la civilisation dans le Nord de la France, *MCDMHPC* IV.2 (1943): *RBPh* XXIII (1944), 163–85: *RN* XXIX (1947), 207: *Gallia* V (1947), 434; IX (1951), 78; XII (1954), 136; XXV (1967), 200; XXIX (1971), 228: Dérolez, 505–33: Cornette, A., Arras et sa banlieue, *RN livre géographique* IX (1960): Leduque, *Atrébatie,* 45 sq: Fossier, 122: *Septentrion* I (1970), 123, 135.

ARTENAY Lug., *Carnutes* (Loiret, F) 48°05′N, 1°53′E
'Auvilliers', Gallo-Roman foundations, small finds, terra sigillata: coin hoard, Valerian to Aurelian.
Rev Soc arch Orl V, 45; XI.1 (1953), 156: Potron, I, 251.

ARUN, river Brit., *Regni* (West Sussex, GB)
TRISANTONA FLVMEN.
Ptol. II,3,3 (Τρισάντωνος ποταμοῦ ἐκβολαί).
PNRB 476–7.
(The river Arun was formerly known as the Tarrant).

ARVII

Ptol. II, 8, 7 ('Αρούιοι).
Tribe of Lugdunensis with capital at VAGORITVM (?La Bazouge-de-Chémeré, *qv*). Amalgamated under the late Empire with the CIVITAS DIABLINTVM (*cf* AVLERCI).

ASCOUX Lug., *Carnutes* (Loiret, F) 48°07'N, 2°15'E
Gallo-Roman foundations.
Nouel, 20, 32.

ASH Brit., *Cantiaci* (Kent, GB) 51°21'N, 0°19'E
Corridor villa; bathhouse.
KAR XX (1970), 13–20: *VCH Kent* III (1932), 103–4.

ASHTEAD Brit., ?*Regni* or *Atrebates* (Surrey, GB) 51°20'N, 0°19'W
Corridor villa, hypocausts, detached bath-building, coins of 1st and 2nd c; tilery. Built 1st c, abandoned *c* AD 200.
SyAC XXXVII (1927), 144–63; XXXVIII (1929), 1–17, 132–48.

ASNIÈRES Lug., *Lexovii* (Eure, F) 49°12'N, 0°23'E
'Chenappeville', large bath-building, lead conduit, hypocaust.
Gallia XX (1962), 423.

ASNIÈRES Lug., *Parisii* (Hauts-de-Seine, F) 48°55'N, 2°17'E
Early-Empire foundations, sculptures, pottery; Late-Empire cemetery.
Duval, 1961, 231: Roblin, 1971, 70, 183, 260.

*****ASSAINVILLERS** Bel., *Ambiani* (Somme, F) 49°37'N, 2°36'E
'Salle du Bois de l'Épinette', temple.
Agache, 1972, 321: Agache-Bréart, 1975, 27.

ASTHALL Brit., *Dobunni* (Oxfordshire, GB) 51°48'N, 1°35'W
Minor settlement at crossing of river Windrush on the road from Cirencester to Alchester. Traces of buildings, hearths, pits; pottery of 2nd and 3rd c, coins mostly of 4th c. To the south a small house, bath-building, pottery and coins of 3rd and 4th c.
VCH Oxfordshire I (1939), 319–20, 330: *Oxon* XX (1955), 29–39.

*****ATHIES** Bel., *Viromandui* (Somme, F) 49°51'N, 2°59'E
Large villa, hypocaust, apsidal cellar. Earlier and later remains.
RN XLIX (1967), 715: *Gallia* XXV (1967), 203; XXVII (1969), 231; XXIX (1971), 231: Agache, fig. 432–4: Leduque, *Ambianie*, 138: Agache-Bréart, 1975, 27.

ATHIS-MONS Lug., *Parisii* (Essonne, F) 48°43'N, 2°24'E
Gallo-Roman foundations.
Duval, 1961, 227: Roblin, 1971, 71, 181, 269.

ATREBATES Bel.
Caes. *BG* II, 4, 16, 23; IV, 21; V, 46; VII, 75; VIII, 7, 47; and further references to *Atrebas* in the singular.
Str. IV,3,5 (C. 194) ('Ατρεβάτιοι).
Liv. *Ep*. CIV.
Plin. *NH* IV, 106.
Ptol. II,9,4 ('Ατριβάτιοι).
SHA Gallienus, 6 (*Atrabatica regna*); *Carinus*, 20 (*ab Atrabatis*).
Not. Dig. Occ. XLII, 40 (*Atrabatis*).
Not. Gall. VI, 6 (*civitas Atrabatum*).
Hier. *Chron. a. Abraham* 2383; *adv. Iovinianum* 2, 21.
Oros. VI, 7, 11; VII, 32.
Fredegar *Chron* II, 45.
Sid. Apoll. *Carmina* V, 212–3.
CIL XIII.1, pp 558–9 (notes); XIII.2, 9158 (milestone of Tongres, *Fines Atrebatiu[m]*).
Belgic tribe, occupying Artois. Oppidum: Étrun. Capital: NEMETACVM, later known as ATRABATIS (Arras, *qv*).

ATREBATES Brit.
Ptol. II,3,12 and 13 ('Ατρεβάτιοι).
It. A. 478.3 (*Galleva Atrebatum*).
Rav. Cos. 106.32 (*Caleba Arbatium*).
PNRB 259–60.
British tribe of Belgic origin, closely connected with the *Atrebates* of Gaul. Extensive territory in the pre-conquest period (west Sussex, west Surrey, Berkshire, Hampshire and east Wiltshire). Under Roman rule the CIVITAS ATREBATVM was more restricted in extent and had its capital at CALLEVA (Silchester, *qv*).
Rivet, 1964, 139–40: Wacher, 1975, 242–3: Frere, 1978, 56–9: Salway, 1981, 43, 47, 56, 91.

ATWORTH Brit., *Belgae* (Wiltshire, GB) 51°23'N, 2°12'W
Villa; winged corridor; bath suite, coins Gallienus to House of Theodosius. Built *c* AD 200, abandoned *c* 400.
WANHM XLIX (1940–42), 46–95 *Ant. J.* XXIII (1943), 148–52.

AUBÉGUIMONT Lug., *Caletes* (Seine-Maritime, F) 49°48'N, 1°40'E
Gallo-Roman foundations, 'La Mare Close'.
Cochet, *S.I.*, 563.

AUDIERNE Lug., *Osismii* (Finistère, F) 48°01'N, 4°34'W
Salt manufacture. Roads towards Douarnenez, Quimper, Pont l'Abbé.
BSAF, 1875, 122–23: *Gallia* XXX (1972), 207.

AVGVSTODVRVM *cf* Bayeux

AVGVSTOMAGVS *cf* Senlis

AVLERCI
Tribal group with four subdivisions.
AVLERCI unspecified:
Caes. *BG* II, 34; III, 29; VII, 4; VIII, 7.
Liv. V, 34.
Oros. VI, 11.
CIL XIII, 610 (*cives Aulercu[s]*).
AVLERCI BRANNOVICES:
Caes. *BG* VII, 75.
Client tribe of the *Aedui* in central Gaul; not on this map.
AVLERCI CENOMANI:

Caes. *BG* VII, 75.
Plin. *NH* IV, 107.
Ptol. II,8,8 (Αὐλίρκιοι οἱ Ἐνομανοί/Κενομάννοι).
Not. Dig. Occ. XLII, 35 (*Ceromannos*).
Not. Gall. III, 3 (*civitas Cenomannorum*).
Conc. Aurel. a. 511 (*episcopus eclesiae Caelomannicae*).
Ven. Fort. *Vita S. Paterni* X (*Cinomannis*).
Greg. Tur. *HF* II, 31; VIII, 39.
CIL XIII.1, p 508.
Included in Lugdunensis (later Lugdunensis III); occupying the district of Maine with capital at VINDINVM or VINDVNVM, later known as CENOMANNIS (Le Mans, *qv*). There were also *Cenomani* in northern Italy (*cf* Pol. II, 17: Liv. V, 35: Str. V, 1, 9 (C. 216): etc).

AVLERCI DIABLINTES:
Caes. *BG* III, 9.
Plin. *NH* IV, 107 (*Diablinti*).
Ptol. II,8,7 (Αὐλίρκιοι οἱ Διαβλίται).
Not. Gall. III, 10 (*civitas Diablintum*).
Oros. VI, 8.
Not. Tiron. LXXXVII, 64 (*Diablentas*).
CIL XIII.1, pp 507–8.
Included in Lugdunensis (later Lugdunensis III) with capital at NOVIODVNVM DIABLINTVM (Jublains, *qv*).

AVLERCI EBVROVICES:
Caes. *BG* III, 17; VII, 75.
Plin. *NH* IV, 107.
Ptol. II,8,9 (Αὐλίρκιοι οἱ Ἐβουρουικοί).
It. A. 384.4: *Tab. Peut.* (*Mediolano Aulercorum*).
Not. Gall. II, 5 (*civitas Ebroicorum*).
Oros. VI, 8.
Not. Tiron. LXXXVII, 62 (*Ebroice*).
Conc. Aurel. a. 511 (*episcopus de Ebroicas*).
CIL XIII.1, 1390 (*civitatis Aulercorum [Ebu]r(ovicum)*); notes, p 510; 3202 (*[Mediol]anensium*).
Included in Lugdunensis (later Lugdunensis II) with capital at MEDIOLANVM AVLERCORVM, later known as EBROICIS (Évreux, *qv*).
Mathière, 1925, *passim*.

AULNAY-LA-RIVIÈRE Lug., *Carnutes* (Loiret, F) 48°12′N, 2°23′E
Gallo-Roman foundations, quern, coins.
Gallia XXI (1963), 408: Nouel, 31.

AULT Bel., *Ambiani* (Somme, F) 50°06′N, 1°27′E
Villa.
Agache, fig. 238: Leduque, *Ambianie*, No. 686, 689: Agache-Bréart, 1975, 28: Agache, 1978, 37, 65, 74, 75, 76, 78, 82, 257, 287, 304.

AVTRICVM *cf* Chartres
AVTVRA FLVMEN *cf* river Eure

AUVILLIERS Lug., *Caletes* (Seine-Maritime, F) 49°46′N, 1°35′E
Gallo-Roman foundations, pottery, 'Le Mont à Caillot'.
Cochet, *S.I.*, 507.

AUXI-LE-CHÂTEAU Bel., *Ambiani* (Pas-de-Calais, F) 50°14′N, 2°07′E
Bath-building.
Leduque, *Ambianie*, 130.

AVENTIVS FLVMEN *cf* river Ewenny

AVESNES-LE-COMTE Bel., *Atrebates* (Pas-de-Calais, F) 50°17′N, 2°32′E
Important inhumation cemetery, 4th c: villa; ?*vicus*.
Dérolez, 512, 529.

AVION Bel., *Atrebates* (Pas-de-Calais, F) 50°24′N, 2°50′E
Gallo-Roman foundations; 2 *burgi*.
Dérolez, 521.

AVON Lug., *Senones* (Seine-et-Marne, F) 48°24′N, 2°43′E
Temple, small settlement.
Gallia XVII (1959), 280.

AVON, river Brit., *Belgae* (Avon, GB)
ABONA FLVMEN
Rav. Cos. 108.27.
PNRB 239.

AVRANCHES Lug., *Abrincatui* (Manche, F) 48°41′N, 1°21′W
INGENA or LEGEDIA, later ABRINCAS.
Ptol. II,8,8 (Ἰνγένα).
Tab. Peut. (*Legedia*).
Cf ABRINCATVI.
Town, capital of the CIVITAS ABRINCATVORVM, on the road between Rennes and Valognes. Founded 1st c, peak of importance 2nd c, abandoned 4th c. Walls, fort, forum, triumphal arch, aqueducts, houses. Objects of bronze, glass, painted plaster, hypocausts, architectural fragments, querns, statuettes and weights in terracotta. Coins of early and late Empire, one of Domitian, one of Licinius II. Terra sigillata: MARTINI M. VARVCI M. SACERO M. EMO M. NARB. OF. CALVI. SECVNDVS.
de Gerville, 1854, 82: *Bull Mon* XXI (1855), 261–5: *MSAN* XXII (1858), 271–7: *BSAIC* XXIV (1900), 152–6: *BSNÉP* XIII (1905), 156–9: *Rev Av* XVI (1910–11), 132; XXII (1927–9), 261–5; XXVII (1934), 27–37; XXXIX (1961), 23–31; XLI (1964), 63, 64, 86–95; XLIII (1966), 167–79: *Gallia* XXX (1972), 344: Johnson, 1979, 83.

AYLBURTON Brit., *Dobunni* (Gloucestershire, GB) 51°43′N, 2°33′W
Villa; corridor, pottery, *tesserae*, stamped tiles. Occupied 2nd and 3rd c.
JRS XLVI (1956), 141.

AYLESFORD Brit., *Cantiaci* (Kent, GB) 51°19′N, 0°27′E
Tile kiln.
KAR V (1966), 23.

BADBURY Brit., ?*Dobunni* (Wiltshire, GB) 51°32′N, 1°43′W
Villa, courtyard; baths, tessellated floors, pottery of 1st to 4th c.
WANHM LVII (1958), 24–29.

BADBURY Brit., *Durotriges* (Dorset, GB) 50°49′N, 2°03′W
VINDOCLADIA
It. A. 483.5 (*Vindogladia*); 486.14 (*Vindocladia*). Rav. Cos. 106.17 (*Bindogladia*).
PNRB 500.
Minor settlement on the road from Silchester to Dorchester (Dorset), succeeding large pre-Roman hill fort. Pottery, bones, coins dating from Vespasian to Arcadius, but especially AD 250 to 400.
PDNHAS LIV (1932), 87–90; RCHM *Dorset* V, 1975, 60–1.

BAGBER Brit., *Durotriges* (Dorset, GB) 50°47′N, 2°17′W
Pottery kiln, rectangular building, coins mid 2nd to mid 3rd c.
PDNHAS XCV (1974), 93.

BAGUER-MORVAN Lug., *Redones* (Ille-et-Vilaine, F) 48°32′N, 1°47′W
'la Boissière', Gallo-Roman foundations. Tegulae; Lezoux pottery.
Gallia XXXIII (1975), 333.

BAIGNEAUX Lug., *Carnutes* (Eure-et-Loir, F) 48°08′N, 1°49′E
Gallo-Roman foundations, terra sigillata, coins, statuette.
La France illustrée, 1852: Nouel, 16, 22, 29.

BAILLEUL-NEUVILLE Lug., *Caletes* (Seine Maritime, F) 49°48′N, 1°25′E
At 'Les Carrières' Gallo-Roman foundations and other remains.
Cochet, *S.I.*, 542.

BAILLEULVAL Bel., *Atrebates* (Pas-de-Calais, F) 50°13′N, 2°38′E
Kiln.
Dérolez, 516.

BAILLOLET Lug., *Caletes* (Seine-Maritime, F) 49°49′N, 1°25′E
Gallo-Roman foundations; square building 5 m× 5 m, near the Croix des Trois-Frères.
Cochet, *S.I.*, 541.

BAILLY-EN-RIVIÈRE Lug., *Caletes* (Seine-Maritime, F) 49°55′N, 1°20′E
Small settlement; foundations.
Cochet, *S.I.*, 311: *BCASM* XIV, 217.

BAIOCAS *cf* Bayeux

BAIOCASSES
Plin. *NH* IV, 107 (*Bodiocasses*).
Ausonius, *Commemoratio Professorum Burdigalensium* V, 7 (*Bagocassi*).
Not. Dig. Occ. XLII, 34 (*Baiocas*).
Not. Gall. II, 3 (*civitas Baiocassium*).
Not. Tiron. XV, p 142 (*Baiocas*).
Sid. Apoll. *Ep* IV,18,2 (*in praedia Baiocassina*).
Ven. Fort. *Vita S. Paterni* X (*Baiocas*).
Greg. Tur. *Vita S. Martini* II, 53 (*Baiocasinsis civitas*); *HF* V, 26; IX, 13; X, 9 (*Baiocassini*).
CIL XIII. 1, p 496.
Tribe of Lugdunensis (later Lugdunensis II) occupying the Bessin with capital at AVGVSTODVRVM, later known as BAIOCAS (Bayeux, *qv*).

*BAIZIEUX Bel., *Ambiani* (Somme, F) 50°00′N, 2°31′E
Gallo-Roman foundations, small finds.
Vasselle, 296: Agache-Bréart, 1975, 31: Agache, 1978, 446.

BALDOCK Brit., *Catuvellauni* (Hertfordshire, GB) 51°59′N, 0°11′W
Town, occupied 1st c BC to end of the Romano-British period. Large cemetery.
JBAA, 2nd s, XXXVIII (1932), 235–69: *JRS* LIX (1969), 221–22: *Britannia* II (1971), 269; IV (1973), 298: Rodwell-Rowley, 1975, 125–29.

BANNEVILLE-LA-CAMPAGNE Lug., *Viducasses* (Calvados, F) 49°10′N, 0°13′W
Small villa, 'Chemin Saulnier', on the road between Escoville and Soliers. Tiles, green slates, oyster shells; a coin of Constantine. 4th c.
BSAN XLIX (1942–45), 428.

BANVILLE Lug., *Viducasses* (Calvados, F) 49°19′N, 0°29°W
?GRANNONA (*qv*).
Not. Dig. Occ. XXXVII, 14 and 23.
Fort on the left bank of the river Seule defended by two ramparts. Tiles, fragments of mosaic, large defensive hooks fixed in the rock. Burials with large bronze rings. 2nd c.
de Caumont, 1831, 3–5–17: 1857, 541–43: *BSAN* XXIX (1913–14), 222.

BANWELL Brit., *Belgae* (Avon, GB) 51°20′N, 2°52′W
Villa; mosaic, hypocausts, bath-building. Pottery late 3rd to mid 4th c, when abandoned.
VCH Somerset I (1906), 307–08: *JRS* LVIII (1968), 199: Rainey, 1973, 21.

BAONS-LE-COMTE Lug., *Caletes* (Seine-Maritime, F) 49°38′N, 0°46′E
Cremation cemetery on the Yvetot–Varengeville road.
BCASM XV, 35–37.

BAPAUME Bel., *Atrebates* (Pas-de-Calais, F) 50°06′N, 2°51′E
Major settlement.
Nimal, P., Remarques sur les origines antiques de Bapaume, *RN* XLVII (1965), 635; *Naissance de Bapaume*, Bapaume, 1966; *Bapaume Antique*, Arras, 1967.

BARDOUVILLE Lug., *Lexovii* (Seine-Maritime, F) 49°25′N, 0°56′E
At Hameau de Beaulieu, villa *c* 3rd to 4th c.
BSNÉP XXXVIII, fasc 4, 155: *Gallia* XXIV (1966), 260.

BARDOWN Brit., *Regni* (East Sussex, GB) 51°03′N, 0°23′E
Iron mine, wooden buildings, furnaces, pottery, wheels, tiles imprinted CL BR *c* 150–250.
Cleere, H. F., *The Romano-British Industrial Site at Bardown*, in *SxAC Occ Pap* (1970), *passim*.

BARENTIN Lug., *Caletes* (Seine-Maritime, F) 49°33′N, 0°57′E

Minor settlement, foundations: two cremation cemeteries, one by the church, the other at 'La Forterelle'; 4th c cemetery in the Rue des Martyrs.

Cochet, *S.I.*, 184: *BSNÉP* XXIV, 125–26.

*****BARLEUX** Bel., *Ambiani* (Somme, F) 49°53′N, 2°53′E

Temple – sanctuary – with important associated buildings.

Agache, *BSAP* 1972, 321: Agache-Bréart, 1975, 31: Agache, 1978, 396, 400, 402.

BARNSLEY PARK Brit., *Dobunni* (Gloucestershire, GB) 51°45′N, 1°53′W

Villa, baths, field system. Built 2nd c, flourished especially between AD 330 and 375.

TBGAS LXXXVI (1967), 74–83: *Britannia* I (1970), 293; II (1971), 274; III (1972), 338; IV (1973), 307, 331; V (1974), 446; VI (1975), 271; VII (1976), 352; IX (1978), 455; X (1979), 318; XI (1980), 382–4: RCHM, *Gloucestershire Cotswolds*, 1976, 9–11; McWhirr, 1981, 89–90.

BARON-SUR-ODON Lug., *Viducasses* (Calvados, F) 49°08′N, 0°29′W

Temple at 'Le Mesnil'. Decagonal structure within a double temenos. Cenotaph of 2nd c. Five hundred bronze votive rings; early Gallo-Roman fibula; coins of 1st, 2nd, 3rd and especially 4th c, up to the sons of Constantine; pottery, glass, small objects of bone, wall paintings, fragments of mosaic.

de Caumont, 1846, 113: *BSAN* LII (1952), 263–4, 310; LIV (1961–2), 766–7; LXIII (1969), 436–41: *RÉA* LVI (1954), 410: *BSHAO* XVII (1959), 325–6: Lantier, 1966, no 9182: *Gallia* XVII (1959), 325–6; XXX (1972), 333–5; XXXV (1977), 75–88.

BARROW HILL, WEST MERSEA Brit., *Trinovantes* (Essex, GB) 51°48′N, 0°56′E

Tumulus; diameter at base 33·8 m; height 6·7 m. Tiled burial pit; lead casket containing a green glass urn, pottery.

VCH Essex III (1963), 159–61.

BATH Brit., *Belgae* (Avon, GB) 51°23′N, 2°21′W

AQVAE SVLIS

Ptol. II.3.13 (Ὕδατα Θερμά).

It. A. 486.3 (*Aquis Sulis*).

RIB 138–178.

PNRB 255–6.

Town, based on pre-Roman spa, shrine of the Celtic goddess SVLIS. Developed into Romano-British town under the Flavians. Temple of SVLIS MINERVA, three baths, houses, walls built late 2nd c enclosing 9·5 ha. Cemeteries. Stone quarry to southeast. Bath was taken by the Saxons after the battle of Dyrham in AD 577.

Ross, 1967, 30 etc.: Cunliffe, B., *Roman Bath*, Oxford, 1969: Rainey, 1973, 22: Rodwell-Rowley, 1975, 131–8: *Britannia* VII (1976), 1–32; X (1979), 101–7; XII (1981), 375.

BAUDREVILLE Lug., *Carnutes* (Eure-et-Loir, F) 48°20′N, 1°54′E

Major settlement at 'Ormeville, La Mône', important crossroads in pre- and Gallo-Roman periods; staging post, inn, temple, coin hoard, querns, mosaics.

Air photograph and excavations, Jalmain: *Gallia* XXX (1972), 316.

BAULNE Lug., *Carnutes* (Essonne, F) 48°30′N, 2°21′E

Cemetery.

Gallia VII (1949), 111.

BAUX-STE-CROIX (Les) Lug., *Aulerci Eburovices* (Eure, F) 48°58′N, 1°06′E

Temple.

Vesly, 9.

BAYEUX Lug., *Baiocasses* (Calvados, F) 49°15′N, 0°42′W

AVGVSTODVRVM, later BAIOCAS

Tab. Peut. (*Augustoduro*).

CIL XIII, 3161, 8977–88.

Cf BAIOCASSES.

Town with roads towards Vieux, Lisieux and Jublains and an important road towards le Bac du Port on the river Orne. Baths, aqueduct, milestones; temple of Belinus beneath the cathedral; cemetery at St Floxel with cinerary urns. Many small finds including an oculist's stamp; coins from 1st to 4th c up to Arcadius.

MSAN I (1824), 17–49, 484–9; II (1825), 146–56; V (1829–30), 331–5; VI (1831–3), 37–8; XII (1841), 430–33; XXVIII (1873), 75–83: de Caumont, 1857, 450–61: *BSAN* XI (1883), 609–12: Blanchet, *Enceintes*, 36–7: Espérandieu IV (1911), 3045, 3049, 3051: *Gallia* II (1944), 275: Grenier, *Manuel* IV.1 (1960), 355–61.

*****BAYONVILLERS** Bel., *Ambiani* (Somme, F) 49°52′N, 2°38′E

Major settlement, 'Chemin vert'; coin hoard from Aelius to Alexander Serverus.

BSAP XXVIII, 136: Agache, fig. 600: Agache-Bréart, 1975, 31–2.

BAZINGHEM Bel., *Morini* (Pas-de-Calais, F) 50°50′N, 1°40′E

Tumulus with Gallo-Roman material.

Gallia XXV (1967), 200.

BAZOCHES-EN-DUNOIS Lug., *Carnutes* (Eure-et-Loir, F) 48°06′N, 1°34′E

Villa (transformed into a cemetery in Merovingian times).

Gallia XXX (1972), 316.

BAZOCHES-LES-GALLERANDES Lug., *Carnutes* (Loiret, F) 48°10′N, 2°03′E

Gallo-Roman foundations, cemetery; quern, coins.

Nouel, 20, 22, 27, 31, 35.

BAZOCHES-LES-HAUTES Lug., *Carnutes* (Eure- et-Loir, F) 48°09′N, 1°48′E

Gallo-Roman foundations, cemetery; head of Venus.

Nouel, 6, 15, 16, 27, 29, 30, 38.

BAZOUGE-DE-CHÉMÉRÉ (La) Lug., *Arvii*? (Mayenne, F) 48°00′N, 0°30′W

?VAGORITVM
Ptol. II,8,7 (Οὐαγόριτον).
Cf ARVII.
?Town, capital of the CIVITAS ARVIORVM under the early Empire.
Caesarodunum IX.2 (Paris, 1974), 367–378.

BEAUBEC-LA-ROSIÈRE Lug., *Caletes* (Seine-Maritime, F) 49°39′N, 1°32′E
At 'Le Vimel' a cemetery, 1st to 3rd c.
Cochet, *S.I.* 575: *BSNÉP* XXIII.

BEAUCE Lug., *Redones* (Ille-et-Vilaine, F) 48°20′N, 1°10′W
'la Salle', pottery kiln.
BSAIV XXXIX (1910), 179–94.

BEAUCHAMPS Bel., *Ambiani* (Somme, F) 50°01′N, 1°31′E
Gallo-Roman foundations beneath the Abbey of Lieu-Dieu; coins, inscriptions.
Leduque, *Ambianie*, 168.

*****BEAUCOURT-EN-SANTERRE** Bel., *Ambiani* (Somme, F) 49°47′N, 2°35′E
Gallo-Roman foundations; tiles.
Vasselle, 296: Agache-Bréart, 1975, 32: Agache, 1978, 423.

BEAUMONT-CUM-MOZE Brit., *Trinovantes* (Essex, GB) 51°54′N, 1°13′E
Salt working; briquetage, pottery.
CAGQB II (1959), 26–27: *VCH Essex* III (1963), 47.

BEAUMONT-LE-ROGER Lug., *Aulerci Eburovices* (Eure, F) 49°05′N, 0°47′E
Temples at 'Côtes de Fontaine l'Abbé, Triage du puits des Essarts' near sources of water.
Grenier, *Manuel* IV.2 (1960), 788–90.

BEAUMONT-SUR-OISE Lug., *Parisii/Veliocasses* (Val d'Oise, F) 49°08′N, 2°17′E
Gallo-Roman foundations at the site of the old bridge; tiles, pottery, coins. On the side of the hill at 'Pres de Thury, Le Trou d'Enfer', foundations, cemetery, pottery, coins, a roadway.
Duval, 1961, 42, 95: *BAVF* VI (1970), 105: *Gallia* XXX (1972), 307.

BEAUPORT PARK Brit., *Regni* (East Sussex, GB) 50°54′N, 0°33′E
Iron mine. Furnaces, bath-house, coins, pottery, iron statuette, *terra sigillata*, tiles stamped CL BR.
Straker, 1931, 330–7: *SxAC* CVII (1969), 111: *Britannia* X (1979), 139–156.

BEAUQUESNE Bel., *Ambiani* (Somme, F) 50°05′N, 2°24′E
Cemetery
Leduque, *Ambianie*, 133: Agache, 1978, 65.

BEAUVAIS Bel., *Bellovaci* (Oise, F) 49°26′N, 2°05′E
CAESAROMAGVS, later BELLOVACIS or BELVACVS
Ptol. II,9,4 (Καισαρόμαγος).
It. A. 380.3, 384.8 (*Caesaromago*).
Tab. Peut. (*Casaromago*).
CIL XIII, 9026 (milestone on the Troyes-Amiens road).
Cf BELLOVACI.
Town and road junction, capital of the CIVITAS BELLOVACORVM (but probably not identical with BRATVSPANTIVM: cf Vendeuil-Caply). Roman house, destroyed end of 3rd c; cellars at the foot of the rampart (living quarters); semi-circular portico (perhaps a central room, fountain or public garden, end of 3rd c). Beneath the cathedral, superimposed houses and a cellar. Coin hoards. Baths at St Étienne's church; amphitheatre; cemetery (sarcophagi). Late-Empire enclosure (re-used) designed with the systematic rules of a fort (10 ha); towers, gateway. At Mont Capron, temple discovered in the 17th c. Near Beauvais, at the abbey of St Lucien, a cemetery, coins. At 'Les Longs Prés', a cemetery. La Tène I-II pottery in a Gallo-Roman stone-cutters' workshop. Musée Départmental de l'Oise.
Labande, L.-H., *Histoire de Beauvais*, Paris, 1896: *Congrès archéologique de France, 1905* (1906), 336–391: Blanchet, *Enceintes*, 116–20: *BACTH*, 1915, 3–39: Grenier, *Manuel* I (1931), 427; III.1 (1958), 416: Fauqueux, H., *Beauvais et son histoire*, B., 1938: Duval, 1961, 131, 285, 286: *Hommages Grenier* (Brussels, 1962), 1134–50: *Ogam* XV (1963), 47–62: *RN* no 195 (Oct-Dec 1967), 637: *La Revue de Louvre* XVIII (1968), 117–120: *Britannia* IV (1973), 210–23: *Gallia* VII (1949), 112; IX (1951), 82; XII (1954), 144; XV (1957), 164; XVII (1959), 282; XIX (1961), 301, XXI (1963), 367–9; XXV (1967), 196; XXVII (1969), 231; XXIX (1971), 225: Roblin, 1978, 213–20.

BEDWYN BRAIL Brit., *Atrebates* (Wiltshire, GB) 51°22′N, 1°36′W
Villa; pottery of late 2nd, 3rd and 4th c.
WANHM XLVIII (1938), 318–20: *VCH Wiltshire* I (1957), 73.

BELBEUF-SAINT-ADRIEN Lug., *Veliocasses* (Seine-Maritime, F) 49°24′N, 1°08′E
At Saint Adrien, beside the *route nationale*, near the mouth of the Becquet, foundations of a house.
BCASM XV, 111; XVI, 370.

BELERIVM PROMONTORIVM cf Land's End

BELGAE
Ptol. II,3,13 (Βέλγαι).
It. A. 478.2, 483.2 (*Venta Belgarum*), 486.11 (*Venta Velgarum*).
Rav. Cos. 106.18 (*Venta Velgarom*).
RIB 2222.
PNRB 267, 492.
British *civitas*, created by the Romans from several *pagi* whose pre-Roman status is unknown. Its territory stretched from Hampshire through Wiltshire into Avon, with capital at VENTA BELGARVM (Winchester, qv). According to Ptolemy it also included AQVAE SVLIS (Bath) and ISCALIS (?Charterhouse).
Rivet, 1964, 140–1.

BELGIUM

A geographic and ethnic concept which includes the peoples situated between Seine and Somme – *Ambiani, Bellovaci, Meldi, Silvanectes, Veliocasses* and perhaps *Atrebates* and *Viromandui*.

Hawkes, C. F. C., *New thoughts on the Belgae* in *Ant* XLII (1968), 6–16.

BELLENCOMBRE Lug., *Caletes* (Seine-Maritime, F) 49°43′N, 1°13′E

Gallo-Roman foundations; coins in the field adjoining the château.

Cochet, *S.I.,* 269.

BELLOVACI

Caes. *BG* II,4,5,10,13,14; V,46; VII,59,75,90; VIII,6,7,12,14–17, 20–23, 38.

Cic. *ad F* VIII,1,4.

Liv. *Ep* CVIII.

Str. IV,3,5 (C. 194); 4,3 (C. 196); 5,11 (C. 208) (Βελλόακοι).

Plin. *NH* IV, 106.

Ptol. II,9,4 (Βελλούακοι).

Paneg. Const. Caes. XXI.

Not. Gall. VI, 10 (*civitas Bellovacorum*).

Oros. VI, 7, 11 (*Bellovagui*).

Nôt. Tiron.

CIL XII, 1922 (*civi Bellova[co]*); XIII.1, 611 (*c(ivis) Bel[l(ovaci)]*); notes, p 547.

Head, 9.

Belgic tribe, occupying Beauvaisis. Oppida at Bailleul/Thérain, *Bratuspantium* (?Vendeuil-Caply, *qv*), Catenoy; capital at CAESAROMAGVS, later known as BELLOVACIS (Beauvais, *qv*).

Recueil des antiquités bellovaques, Paris/Beauvais, 1860: *Geog J* LXI (1923), 44: *BSNAF*, 1959, 263–81; 1969 (1971), 295–320: *Celticum* III (1961), 37–56; XV (1966), 127–138: Matherat, G., La 1er campagne de César contre les B., *Mélanges Grenier*, Brussels, 1962, 1134–50: *Actes du 91e Congrès des Sociétés Savantes*, Rennes, 1966, 203–38.

BELLOY-EN-SANTERRE Bel., *Ambiani* (Somme, F) 49°53′N, 2°51′E

Small villa.

Agache-Bréart, 1975, 33.

*BELLOY-SAINT-LEONARD Bel., *Ambiani* (Somme, F) 49°44′N, 1°55′E

Temple.

Agache-Bréart, 1975, 33.

BELLOY-SUR-SOMME Bel., *Ambiani* (Somme, F) 50°06′N, 1°32′E

At the hamlet of St-Accart, Gallo-Roman foundations, sarcophagi, tiles.

Agache, fig. 468, 555 (2 sites): Leduque, *Ambianie*, 77: Agache-Bréart, 1975, 33–34: Agache, 1978, 89, 316.

BENENDEN Brit., *Cantiaci* (Kent, GB) 51°04′N, 0°34′E

Iron mine, furnaces, tiles.

Arch Cant LXXVIII (1963), 189.

BÉNOUVILLE Lug., *Viducasses* (Calvados, F) 49°15′N, 0°16′W

'le Catillon', fort? Bath-house, villa, cemetery with cinerary urns, pottery, glass, coins: 2nd c.

MSAN XIX (1852), 485–500: *BSAN* XXIX (1913–14), 213.

BERKHAMPSTED Brit., *Catuvellauni* (Hertfordshire, GB) 51°46′N, 0°35′W

On Northchurch Common a corridor villa built in the Antonine period, with later development.

Britannia V (1974), 438: *Herts Arch* IV (1976), 1–135: Todd, 1978, 33–58 *passim*.

BERNIÈRES-SUR-MER Lug., *Viducasses* (Calvados, F) 49°20′N, 0°25′W

Small villa, 'Tombettes de St Ursin'; on the shore on the road betwen Bayeux and le Bac du Port, and a side road to Bény. Small finds, pottery both coarse and terra sigillata VXTVLIM, ENTILIS: 2nd c.

de Caumont, 1846, 375: *BSAN* XXIX (1913–14), 224; XXXII (1917), 275; XXXV (1921–23), 420–29; LIV (1959), 419–20.

BERNY *cf* Pierrefonds

*BERTANGLES Bel., *Ambiani* (Somme, F) 49°58′N, 2°17′E

Cemetery; tiles.

Agache, fig. 23, 53. Leduque, *Ambianie*, 131: Vasselle, 298: Agache-Bréart, 1975, 34: Agache, 1978, 109.

BERTHOUVILLE Lug., *Lexovii* (Eure, F) 49°11′N, 0°38′E

CANETONVM

CIL XIII, 3183.

Two temples in a sanctuary at 'Le Trésor'; abandoned under Tetricus. Courtyard, theatre, deposit of silverware probably from the temple of Mercury Canetonnensis (Paris, Bibliothèque Nationale).

Babelon, E., *Le trésor d'argenterie de Berthouville, près Bernay* (*Eure*), Paris, 1916: Grenier, *Manuel* III.2 (1958), 956–8; IV.2 (1960), 768–77: Duval, 1961, 185, 187.

BERVILLE-SUR-SEINE Lug., *Lexovii* (Seine-Maritime, F) 49°28′N, 0°55′E

Gallo-Roman foundations at Carrière Beaudelin.

Deglatigny, 1925, 27.

BESNEVILLE Lug., *Unelli* (Manche, F) 49°23′N, 1°37′W

Gallo-Roman foundations, bricks, tiles, iron slag, weights. Coins of Hadrian, Gallienus, Postumus.

Et G et H Manche, 1854, 87: *BSAIC* XXIV (1900), 71.

BÉTHUNE Bel., *Atrebates* (Pas-de-Calais, F) 50°32′N, 2°38′E

Small settlement; capital of the PAGVS BETHVNIA.

*BEUVRAIGNES Bel., *Ambiani* (Somme, F) 49°39′N, 2°46′E

Gallo-Roman foundations; kilns, cellars, wells, plaster, small finds.

Leduque, *Ambianie*, 932: Vasselle, 298: Agache-Bréart, 1975, 35: Agache, 1978, 253, 260, 267.

BEUZEC-CAP-SIZUN Lug., *Osismii* (Finistère, F) 48°05′N, 4°31′W

'Kerzeon', villa; hypocaust.
BSAF, 1970, 29.

BÉVILLE-LE-COMTE Lug., *Carnutes* (Eure-et-Loir, F) 48°26′N, 1°43′E
Villas at 'La Villeneuve'.
Air photograph, Jalmain: Nouel, 20.

BEXLEY Brit., *Cantiaci* (Kent, GB) 51°26′N, 0°09′E
Pottery kiln, 2nd c.
Arch Cant LXVIII (1954), 167–83.

BÉZANCOURT Lug., *Veliocasses* (Seine-Maritime, F) 49°27′N, 1°37′E
Gallo-Roman foundations at 'Le Catelier'; imperial coins.
Cochet, *S.I.*, 585.

BIACHE-ST-VAAST Bel., *Atrebates* (Pas-de-Calais, F) 50°18′N, 2°57′E
Temple, coins; two villas, hypocausts.
Jelski, 139: Leduque, *Atrébatie*, 81.

BIERRE Lug., *Esuvii* or *Sagii* (Orne, F) 48°50′N, 0°02′W
Small settlement, possible fort, Iron Age site reoccupied in Roman times.
MSAN IX (1835), 460.

BIGNOR Brit., *Regni* (West Sussex, GB) 50°55′N, 0°35′W
Villa; large courtyard, close to Stane Street between London and Chichester. Fine mosaics. Occupied 2nd to 4th c.
Winbolt, S. E., and Herbert, G., *The Roman Villa at Bignor, Sussex*, Chichester, 1934: *VCH Sussex* III (1935), 22: *JRS* LIII (1963), 155–56: Rainey, 1973, 23: *Britannia* VI (1975), 118–32; XIII (1982), 135–95.

BILLERICAY Brit., *Trinovantes* (Essex, GB) 51°38′N, 0°26′E
Small settlement; potters' workshops, iron furnaces; cemetery.
VCH Essex III (1963), 48; Rivet, 1964, 147: *Britannia* III (1972), 331: Rodwell-Rowley, 1975, 85–101, *passim*.

BILLY-MONTIGNY Bel., *Atrebates* (Pas-de-Calais, F) 50°25′N, 2°54′E
Cemetery, large, 1st c: ?small settlement.
Dérolez, 512.

BINIC Lug., *Osismii* (Côtes du Nord, F) 48°36′N, 2°50′W
'la Blanche', Gallo-Roman foundations. Two hundred small bronze coins of Gallienus, Valerianus etc; more than 1000 small bronze coins composing a series of the principal emperors or usurpers commencing with Trebonianus Gallus and finishing with Claudius Gothicus.
Bull Soc Brest V (1867), appendix 1–183.

BIRDCOMBE *cf* Wraxall

BISHOPSBOURNE Brit., *Cantiaci* (Kent, GB) 51°14′N, 1°07′E
Three tumuli in line, each with a stone cist containing cinders and burnt bone, small bronze ornaments, pottery, tiles, glass.
VCH Kent III (1932), 146–7.

BISHOP'S STORTFORD Brit., *Trinovantes* (Hertfordshire, GB) 51°53′N, 0°11′E
Major settlement on the road between Braughing and Colchester. Potters' workshops, foundations, cemetery.
JRS XLVII (1957), 219: Rivet, 1964, 147: *VCH Herts* IV (1914), 150.

BISHOPSTONE Brit., ?*Dobunni* (Wiltshire, GB) 51°32′N, 1°38′W
Villa; mosaics, hypocausts.
VCH Wiltshire I (1957), 42: Rainey, 1973, 25.

BISHOP'S WALTHAM Brit., *Belgae* (Hampshire, GB) 50°56′N, 1°14′W
Tile kiln.
VCH Hampshire I (1900), 309: *PHFC* XXII (1961), 22.

BITTERNE Brit., *Belgae* (Hampshire, GB) 50°55′N, 1°23′W
?CLAVSENTVM
It. A. 478.1.
RIB 97, 2222–2228.
PNRB 308–9.
Town and port near the mouth of the river Itchen. Timber buildings, Claudian, rebuilt in stone from *c* AD 180. Stone wall of 4th c. Destroyed *c* AD 400. Quays, pottery, many small finds. Milestones built into the wall – Gordian (AD 238–44), Gallus and Volusian (AD 251–3), Tetricus (AD 270–3), Aurelian (AD 273–5) and one unnamed emperor.
Ant J XXVII (1947), 151–71: Cotton, M. A., and Gathercole, P. W., *Excavations at Clausentum*, London, 1958: Ross, 1967, 206: Sedgley, 1975, 20–21.

BIVILLE-SUR-MER Lug., *Caletes* (Seine-Maritime, F) 49°59′N, 1°15′E
Gallo-Roman foundations, tiles, pottery at 'Le Vieux Biville'.
Cochet, *S.I.*, 314.

BLACKBOY PITS *cf* St Stephen

BLAEN-CWM-BACH Brit., *Silures* (West Glamorgan, GB) 51°40′N, 3°44′W
Temporary camp, area 24.4 ha. Perhaps constructed during the campaign of Julius Frontinus *c* AD 75:
JRS XLIX (1959), 102: RCAHM (Wales), *Glamorgan* I, ii, 1976, 99–101.

BLANDY Lug., *Carnutes* (Essonne, F) 48°19′N, 2°15′E
Gallo-Roman foundations at 'Vers Audeville, les Terres-Noires'.
Air photograph, Jalmain: *Gallia* V (1947), 440.

BLANGY-SUR-BRESLE Lug., *Caletes* (Seine-Maritime, F) 49°56′N, 1°37′E
Villa at 'La Planche du Lieutenant'. Near 'Moulin aux Armures', cemetery, pottery, coins.
Cochet, *S.I.*, 545: Leduque, *Ambianie*, 168, n. 774–75.

*****BLANGY-TRONVILLE** Bel., *Ambiani* (Somme,

F) 49°53′N, 2°25′E
Villa, on site of earlier enclosure.
Agache-Bréart, 1975, 35: Agache, 1978, 138, 152, 168, 172, 287, 370.

BLASHENWELL Brit., *Durotriges* (Dorset, GB) 50°37′N, 2°04′W
Shale working. Waste cores of shale, pottery, inhumation burials. Pre-Roman, Romano-British and post-Roman occupation.
RCHM *Dorset* II.3 (1970), 599.

BLESTIVM *cf* ?Monmouth

BODIAM Brit., *Regni* (East Sussex, GB) 51°01′N, 0°33′E
Port, for exporting iron: Romano-British bronze figurine, terra sigillata, tiles imprinted CL BR, glass.
SxAC CIV (1966), 88–102.

BODILIS Lug., *Osismii* (Finistère, F) 48°32′N, 4°07′W
'Spernen' = 'l'Epinaie', Gallo-Roman foundations along a branch of the road between Carhaix and Plouguerneau; coarse pottery and terra sigillata, 1st, 3rd and 4th c.
BSAF, 1967, 1–3.

BOËSSÉ-LE-SEC Lug., *Aulerci Cenomani* (Sarthe, F) 48°08′N, 0°34′E
Villa, 'La Tercinnerie'.
Bouton.

BOIS-AU-MONT Bel., *Atrebates* (Pas-de-Calais, F) 50°12′N, 2°47′E
Temple at 'Le Bois-Potel' southwest of the village.
Agache, *BSAP*, 1972, 321.

BOIS-L'ABBÉ *cf* Eu

BOISMONT Bel., *Ambiani* (Somme, F) 50°09′N, 1°41′E
Minor settlement at 'Fond des Mautoires'; temple, sculptures.
Prarond, E., *Histoire de cinq villes et de 300 villages* III (Abbeville, 1863): Vasselle, 298: Agache, fig. 14, 569–70; *BSAP,* 1972, 321: Leduque, *Ambianie*, 84: Agache-Bréart, 1975, 35.

BOLAZEC Lug., *Osismii* (Finistère, F) 48°27′N, 3°35′W
'Beuzidel', villa; 1st, 3rd and 4th c.
BSAF, 1971, 23–24.

BOLBEC Lug., *Caletes* (Seine-Maritime, F) 49°34′N, 0°27′E
Cemetery, cremations of 1st to 3rd c, at the bottom of the valley of la Fontaine-Martel: another at the hamlet of Roncherolles.
Cochet, *S.I.*, 390.

BOLLEVILLE Lug., *Unelli* (Manche, F) 49°19′N, 1°34′W
'la Terre du Buisson', small villa beside the road from Cherbourg to Coutances. Foundations, small bronze coins of the early Empire.
Et G et H Manche, 1854, 89: *BSAIC* XXIV (1900), 105.

BONONIA *cf* Boulogne-sur-Mer

BOOS Lug., *Veliocasses* (Seine-Maritime, F) 49°24′N, 1°13′E
Villa, corridor, with hypocaust and baths at 'les bois des Marettes'. In the grassland of the 'Porte Rouge' near the 'Bois Flahaut' a rustic villa. Gallo-Roman foundations at 'Le Faulx'. *Cf* Franquevillette.
Cochet, *S.I.,* 211: Vesly, *Amis des monuments rouennais*, 1907, 41; *BCASM* XIV, 178; XV, 112–13, 117, 119: *BSNÉP* XVII (1909), 32.

BORDEAUX-SAINT-CLAIR Lug., *Caletes* (Seine-Maritime, F) 49°42′N, 0°15′E
Villa at 'Château Gaillard' near the 'Bois des Loges'.
Cochet, *S.I.*, 355; *Revue de Rouen*, Jan 1843.
Corridor villa 100 m long in plain of 'Bordeaux' towards the 'Petit Val'.
Cochet, *S.I.*, 356.

BOSCHYONS Lug., *Veliocasses* (Seine-Maritime, F) 49°28′N, 1°39′E
Gallo-Roman foundations.
Cochet, *S.I.*, 566.

BOSC-LE-HARD Lug., *Caletes* (Seine-Maritime, F) 49°38′N, 1°11′E
Iron workings.
Cochet, *S.I.*, 274.

BOSCOMBE DOWN Brit., *Belgae* (Wiltshire, GB) 51°08′N, 1°42′W
Villa, corridor; occupied late 3rd and 4th c.
Britannia II (1971), 281.

BOSVILLE Lug., *Caletes* (Seine-Maritime, F) 49°45′N, 0°42′E
Gallo-Roman foundations.
Cochet, *S.I.*, 454.

BOTLEY Brit., *Belgae* (Hampshire, GB) 50°56′N, 1°16′W
Tile kiln, late 1st c.
PHFC XXII (1961), 22–24.

BOUCÉ Lug., *Esuvii* or *Sagii* (Orne, F) 48°38′N, 0°05′W
?Fort, 'camp de Feuillet or Fouillet'.
MSAN IX (1835), 445–46.

BOUILLANCOURT-EN-SÉRY Bel., *Ambiani* (Somme, F) 49°58′N, 1°38′E
Gallo-Roman foundations, 'Le Château-Bureau'. Abbey of Sery, mosaics, pottery, tiles.
Leduque, *Ambianie*, 167: Vasselle, 298: Agache-Bréart, 1975, 36: Agache, 1978, 41, 215, 452.

BOUILLIE (La) Lug., *Coriosolites* (Côtes du Nord, F) 48°35′N, 2°24′W
The foundations of several Gallo-Roman buildings near the crossing of the roads between Corseul and Erquy and between Carhaix and Aletum (Saint-Servan). Three coin hoards – Tetricus, Maximinus and following, Constantine. Potsherds; statuette of Cupid and a lion in bronze.
ACN , 1863, 11–12: *BSCN* XLVII (1909), 39.

BOULOGNE-LA-GRASSE Bel., *Ambiani* (Oise, F) 49°36′N, 2°43′E
Cemetery, incinerations with tiles, pottery, arms, coins.

Leduque, *Ambianie*, 297.

BOULOGNE-SUR-MER Bel., *Morini* (Pas-de-Calais, F) 50°43′N, 1°37′E
GESORIACVM, later BONONIA
Mela III, 23 (*Gesoriacum*).
Plin. *NH* IV, 102 (*a Gesoriaco Morinorum*); IV, 106 (*pago qui Gesoriacus vocatur*).
Suet. *Claudius*, 17 (*Gesoriacum*).
Ptol. II,9,1 (Γησοριακόν).
It. A. 356.5, 363.2 (*Gesoriaco*), 376.2 (*a portu Gesoricensi*), 463.4 (*a Gessoriaco*).
It. M. 496.4 (*a portu Gesoriagensi*).
Tab. Peut. (*Gesogiaco quod nunc Bononia*).
Paneg. Const. Caes. 6, 1 (*Gesorigiacensibus muris*); 14, 4 (*Gesorigiacensi litore*).
Origo Constantini imp. II, 4 (*apud Bononiam . . . prius Gesoriacum*).
Eutropius IX, 21 (*Bononiam*).
Amm. Marc. XX, 1,3; 9,9; XXVII, 8,6 (*Bononia*).
Not. Gall. VI, 13 (*civitas Bononiensium*).
Cod. Theod. XI, 16,5 (*Bononiae*).
Iul. Hon. *Cosmographia*, 19 (*Cesuriacum oppidum*).
Sozomenus *Ecc. Hist.* IX, 11,3 (Βονονίαν).
Zosimus VI, 2,2 (Βονωνία).
Beda *Ecc. Hist.* I, 33 (*in Bononia civitate*).
CIL XIII.1.2, p 560.
Epigraphia XXXIII (1972), 70–74 (*Coh I (M)orinor(um) et Cersiacor(um)*).

Town and port at the mouth of the river Liane. Sometimes identified with the *Portus Itius* of Caesar (but *cf* Wissant). Used by Claudius as the base of embarkation for the conquest of Britain, AD 43. Capital under the early Empire of the PAGVS GESORIACENSIS, under the late Empire of the CIVITAS BONONIENSIVM.

Two regions of settlement. At 'La Basse-Ville' (?*Gesoriacum* proper), port, possible fort of the *Classis Britannica*, bridge over the Liane. At 'La Haute-Ville' (?*Bononia* proper) occupation in the Claudian and Flavian periods; houses and temples, 2nd c; massive walls, enclosing *c* 12.5 ha, probably built in the reign of Constantius Chlorus. To the NW, a lighthouse, 'La Tour d'Ordre', built under Caligula (Suetonius, *Gaius* 46) and surviving until 1644. Musée des Beaux Arts et d'Archéologie.

Abot de Businghen, *Recherches historiques concernant la ville de B. et l'ancien comté de ce nom*, Paris, 1822: *MSAB* XVIII (1895–96), 470–81; XIX (1903), 81–156, 332–46: Hamy, E., *B. dans l'Antiquité*, B., 1899: *BSAB* VII (1905), 530–62; VIII (1908–9), 85; XI (1929), 204–8: Blanchet, *Enceintes*, 123–4: Grenier, *Manuel* II.2 (1934), 527–9: Héliot, P., *Histoire de B.*, 1937: *RÉA* XLVI (1944), 299; L (1948), 101–11; LI (1949), 324–6: MCDMHPC V (1947), 25–34: *Hommages à J. Bidez et à F. Cumont* II (1949), 127–33: Leduque, *Boulonnais*, 54–72; *Morinie*, 62, 70: *Gallia Supp* X.1.1 (1957), nos 104–8; XX (1962), 89; XXVII (1969), 226; XXIX (1971), 229: *RA*, 1958.1, 158–82; 1958.2, 40–64; 1966.1, 89–96: *RN* XLII (1960), 363–379; LI (1969), 363; LIII (1971), 669: *Hommages M. Renard* II (1969), 820–27:

Septentrion 1–2 (1969), 32–5, 6 (1976), 5–14: *Latomus* XXXIII (1974), 265–80: Johnston, 1977, 35–8: Johnson, 1979, 84–6.

BOULON Lug., *Viducasses* (Calvados, F) 49°03′N, 0°23′W
Small villa near the church; coins, querns, a female head in a niche.
de Caumont, 1831, 228; 1850, 194–96: Espérandieu IV (1911), 3046.

BOURAY-SUR-JUINE Lug., ?*Carnutes* (Essonne, F) 48°31′N, 2°18′E
Temple, sanctuary of the early Empire.
Duval, 1961, 229: Roblin, 1971, 306.

BOURG-BLANC Lug., *Osismii* (Finistère, F) 48°30′N, 4°30′W
'Kergonc', villa beside the road between Kérilien and le Conquet; a lead sarcophagus, a coin hoard of the 3rd c: 1st to 4th c.
BSAF, 1906, 30–32; 1969, 22–23.

BOURNES GREEN Brit., *Dobunni* (Gloucestershire, GB) 51°44′N, 2°08′W)
Villa; tesselated floors, bricks, pottery, glass, metal, bones; coin hoard, Victorinus and Tetricus predominating. Flourished 3rd c.
JBAA I (1846), 44; II (1847), 324; RCHM, *Gloucestershire Cotswolds*, 1976, 15–16.

BOURNEVILLE Lug., *Lexovii* (Eure, F) 49°23′N, 0°37′E
Gallo-Roman foundations, ditches, inhumations (Beaumont manor).
Gallia XXX (1972), 340.

BOURSEUL Lug., *Coriosolites* (Côtes du Nord, F) 48°29′N, 2°16′W
'la Chapelle Saint Méen', Gallo-Roman foundations, debris.
MSCN I (1883–84), 53.

BOURTON GROUNDS Brit., *Catuvellauni* (Buckinghamshire, GB) 51°59′N, 0°56′W
Romano-Celtic temple and two barrows (Thornborough). Temple rectangular, coins 1st to 5th c. Barrows, late 2nd-c pottery.
Records of Bucks XVI (1953–60), 29–32; XVII (1964), 356–66: Lewis, 1966, *passim*.

BOURTON-ON-THE-WATER Brit., *Dobunni* (Gloucestershire, GB) 51°53′N, 1°45′W
Major settlement at a ford on the river Windrush on the Fosse Way between Cirencester and Leicester. *Mansio* and other buildings, pottery, coins, many small finds. Occupied throughout the Romano-British period.
TBGAS XCI (1972), 92–116: Rodwell-Rowley, 1975, 60: McWhirr, 1981, 62–5.

BOUTAVENT-LA-GRANGE Bel., *Ambiani* (Oise, F) 49°39′N, 1°45′E
Gallo-Roman foundations, bricks, tiles, potsherds, coins.
Leduque, *Ambianie*, 32.

*BOVES Bel., *Ambiani* (Somme, F) 49°51′N, 2°23′E
Gallo-Roman foundations; coins, pottery in the

marsh, 1 capital; in the forest small bronze coins of Constantine and Valens.

Agache, fig. 291, 601: Leduque, *Ambianie*, 200, n 345, 348–49: Vasselle, 300: Agache-Bréart, 1975, 37: Agache, 1978, 22, 73, 166, 262, 374, 389, 438.

BOVIVM *cf* ?Cowbridge

BOX Brit., *Belgae* (Wiltshire, GB) 51°25′N, 2°15′W

Villa, courtyard; mosaics, hypocausts, baths. Painted plaster, pottery; stone head now in Devizes Museum; coins chiefly of late 3rd, early 4th c.

WANHM LXIV (1969), 123–4: Rainey, 1973, 26.

BOXMOOR Brit., *Catuvellauni* (Hertfordshire, GB) 51°44′N, 0°30′W

Villa, winged corridor; mosaics, hypocausts. Five phases of occupation from early 2nd to mid 4th c.

VCH Hertfordshire IV (1914), 154–55: *Britannia* I (1970), 156–62: Rainey, 1973, 27: *Herts Arch* IV (1976), 1–135: Todd, 1978, 33–58 *passim*.

BOXTED Brit., *Cantiaci* (Kent, GB) 51°21′N, 0°39′E

Villa, corridor; mosaics. Romano-Celtic temple.

KAR XVIII (1969–70), 9: *VCH Kent* III (1932), 106–9.

BRACHY Lug., *Caletes* (Seine-Maritime, F) 49°49′N, 0°57′E

Small settlement. Tradition of a vanished town on the plain between Grenville and Brachy.

Cochet, *S.I.*, 286.

BRACQUEMONT Lug., *Caletes* (Seine-Maritime, F) 49°56′N, 1°07′E

(Oppidum of the Caletes. 'Cite de Limes, camp de César'). Numerous Gallo-Roman foundations; vestiges of agricultural buildings. A temple lost beneath the sea.

BSPF XVI (1919), list lxxvii, 186: *MSAN* III (1826), 71: Cochet, *S.I.*, 254: *DAG*: Vesly, 10–11.

BRADFIELD Brit., *Atrebates* (Berkshire, GB) 51°28′N, 1°08′W

Pottery kiln.

Britannia II (1971), 284.

BRADING Brit., *Belgae* (Isle of Wight, GB) 50°40′N, 1°09′W

Large courtyard villa with related field system; mosaics. Occupied 3rd and 4th c.

VCH Hampshire I (1900), 313–15: *Bull Inst Arch Lond* I (1958), 55–74: Rivet, 1969, 43–44; Rainey, 1973, 27.

BRADLEY SPRING *cf* Littleton

BRADWELL-ON-SEA Brit., *Trinovantes* (Essex, GB) 51°44′N, 0°56′E

OTHONA

Not. Dig. Occ. XXVIII, 3 and 13 (*Praepositus numeri Fortensium, Othonae*).

Beda, *Hist. Eccl.* III. 22 (*Ythancaestir*).

PNRB 434–5.

Late fort of the *Litus Saxonicum*. Remains of walls. Chapel constructed in the west gateway by St Cedd *c* AD 654.

RCHM Essex IV (1923), 13: *TEAS* XVII (1926),

198–202: *Arch J* XCVII (1940), 125–50: *VCH Essex* III (1963), 52–5: Cunliffe, 1968, 269: Collingwood-Richmond, 1969, 49: Frere, 1978, 380: Johnson, 1979, *passim*.

BRAINTREE Brit., *Trinovantes* (Essex, GB) 51°52′N, 0°53′E

Major settlement on the road between Braughing and Colchester. Traces of buildings, cemetery.

VCH Essex III (1963), 55: Rivet, 1964, 147: Rodwell-Rowley, 1975, 85–101, *passim*.

BRAMDEAN Brit., *Belgae* (Hampshire, GB) 51°03′N, 1°06′W

Villa, courtyard. Mosaics, coins of late 3rd and 4th c.

VCH Hampshire I (1900), 307–08: *PHFC* XXIX (1972), 41–77: Rainey, 1973, 28.

BRASPARTS Lug., *Osismii* (Finistère, F) 48°18′N, 3°57′W

'Castel Du' = 'le château noir'; Gallo-Roman foundations.

BSAF, 1875, 123.

BRATTON SEYMOUR Brit., *Durotriges* (Somerset, GB) 51°04′N, 2°28′W

Villa. Mosaic; coins and pottery of 3rd and 4th c.

JRS LIX (1969), 227: *Britannia* II (1971), 276; III (1972), 343: Rainey, 1973, 30.

*BRATVSPANTIVM *cf* ? Vendeuil-Caply

BRAUGHING Brit., *Catuvellauni* (Hertfordshire, GB) 51°53′N, 0°01′E

Major settlement on the road between London and *Durovigutum* (Godmanchester, Cambridgeshire). Site of pre-Roman town. Cemetery 1st to 4th c; remains of buildings, roads, Romano-Celtic temple; see also Puckeridge.

Arch XCIII (1949), 32: Rodwell-Rowley, 1975, 139–57.

BRAXELLS FARM *cf* Botley

*BRAY-SUR-SOMME Bel., *Ambiani* (Somme, F) 49°56′N, 2°43′E

Small settlement.

Agache, fig. 507: Fossier I, 197: Vasselle, 302: Agache-Bréart, 1975, 37: Agache, 1978, 20, 287, 339.

BREAGE Brit., *Dumnonii* (Cornwall, GB) 50°06′N, 5°20′W

RIB 2232.

Milestone of Postumus (AD 258–268).

Sedgley, 1975, 23.

BREAN DOWN Brit., ?*Belgae* ?*Durotriges* (Somerset, GB) 51°19′N, 3°01′W

Romano-Celtic temple; two structural periods, both probably Constantinian, perhaps after AD 330. By *c* 370 temple had been ransacked and a smelting furnace constructed in the north annexe. Towards the end of the 4th c a small building was erected presumably as another temple (100 coins later than AD 380) and the old temple rased.

PUBSS VIII. 2 (1958), 106–09; *JRS* XLIX (1959), 129.

BREAUTÉ Lug., *Caletes* (Seine-Maritime, F)

49°38′N, 0°23′E
Cemetery, cremation, 1st to 3rd c, at the hamlet of Givoust.
Cochet, *S.I.*, 382.

BRECON GAER *cf* Y Gaer

BREHAN-LOUDÉAC Lug., *Veneti* (Morbihan, F) 48°04′N, 2°42′W
'Camp de César', cemetery, cinerary urns.
BSPM, 1950, 34; 1972, 42, 104: *Ogam* XI, (1959), 25.

*BREILLY Bel., *Ambiani* (Somme, F) 49°56′N, 2°11′E
Cemetery at 'Montjoie'. ?Sacellum. Shaft of an Ionic column, small finds.
Agache, fig. 57: *BSAP* XV (1886), 98: Leduque, *Ambianie*, 112, n. 318: Agache-Bréart, 1975, 37.

BREST Lug., *Osismii* (Finistère, F) 48°24′N, 4°29′W
?GESOCRIBATE
Tab. Peut.
Town; walls, towers, fortifications of the late Empire probably appertaining to the *Litus Saxonicum*. Two coin hoards of 3rd c.
Rev Prov Ouest 1857, 137–52, 222–38: *Coll Bret* 1863, 135–38: de la Barre de Nanteuil, A.? 1914, 1–13: *BSAF*, 1972, 42–53: Le Gallo, ed., *Histoire de Brest*: Johnson, 1979, 80–81.

BRETTEVILLE L'ORGUEILLEUSE Lug., *Viducasses* (Calvados, F) 49°12′N, 0°30′W
Villa; tiles, filled-up wells NE of the town. Several hundred coins of Postumus, Tetricus, Victorinus, Salonina.
de Caumont, 1846, 269.

BRETTEVILLE-SAINT-LAURENT Lug., *Caletes* (Seine-Maritime, F) 49°46′N, 0°51′E
Theatre, at Beauville-la-Cité. In the neighbourhood, traces of foundations. Tradition of a vanished town (BOSVIE).
Cochet, *S.I.*, 430.

BREVIODVRVM *cf* ?Brionne

BRIARRES-SUR-ESSONE Lug., *Carnutes* (Loiret, F) 48°14′N, 2°25′E
Cemetery. Statue of mother-goddess, jewellery, coins, glass, bronze vessel.
Nouel, 9, 10, 17, 27–29, 33.

BRICQUEBEC Lug., *Unelli* (Manche, F) 49°28′N, 1°37′W
'Camp de Castillon', Gallo-Roman foundations and coin hoard; tiles, bricks, querns, slag, 4 bracelet-ends and two gold ear-rings; a gold coin of Tiberius, and in a bronze vessel 450 small bronzes from Gallienus to Maximian – 1st, 2nd and 3rd c. At a place called 'la Luzerne d'en Haut' a hoard of between 300 and 400 bronze and silver coins from Valerian to Diocletian.
de Gerville, 1854, 90: *MSAN* XXII (1857), CXXI: *BSAIC* XXIV (1900), 73: *BSAN* LIII (1955–6), 197–263.

BRIE-COMTE-ROBERT Lug., *Parisii* (Seine-et-Marne, F) 48°41′N, 2°37′E

Villa. Cinerary urns, terra sigillata, coins 2nd c.
Duval, 1961, 230, 248.

BRIGHTLINGSEA Brit., *Trinovantes* (Essex, GB) 51°48′N, 1°00′E
Villa, corridor; mosaics.
VCH Essex III (1963), 57.

BRIIS-SOUS-FORGE Lug., *Parisii* (Essonne, F) 48°37′N, 2°07′E
Gallo-Roman foundations; coins.
Duval, 1961, 233: Roblin, 1971, 180.

BRIMEUX Bel., *Morini* (Pas-de-Calais, F) 50°27′N, 1°50′E
BRIVA-MAGVS
Gallo-Roman foundations.
Leduque, *Boulonnais*, 30, n 11–12: Delmaire, 1976, 325 etc.

BRIONNE Lug., *Lexovii* (Eure, F) 49°12′N, 0°43′E
?BREVIODVRVM
It. A. 385.2 (*Brevodorum*).
Tab. Peut. (*Brevoduro*).
Major settlement on the Risle.
Le Prévost, 1862, I, 21 and 437: *Ann Norm* I (1960), 80–82.

BRIOT *cf* St Maur-en-Chaussée

BRISLINGTON Brit., *Belgae* (Avon, GB) 51°26′N, 2°33′W
Villa, corridor. Baths, mosaics. Constructed *c* AD 270, destroyed *c* AD 367.
VCH Somerset I (1906), 303–04: *PSANHS* CXVI (1971–72), 78–85: Rainey, 1973, 30.

BRIVA ISARAE *cf* Pontoise

BROADFIELDS Brit., *Regni* (West Sussex, GB) 51°06′N, 0°12′W
Iron workings; 36 furnaces, buildings. Occupation from early 1st to mid 4th c.
Britannia IV (1973), 320; V (1974), 457; VII (1976), 282.

BROCKLEY HILL Brit., *Catuvellauni* (Greater London, GB) 51°38′N, 0°18′W
SVLLONIACIS
It. A. 471.4.
Potters' Stamps, *LVGVDVNVM*.
PNRB 463.
Pottery, on Watling Street between London and St Albans. A very extensive area of kilns, clay-pits and huts; occupation probably pre-Roman and then through from 1st to 4th c. Many small finds include bronze dog, large quantities of pottery; pottery marked SVLLON found at Corbridge is not paralleled here.
TLMAS XVI (1951), 1–23, 201–28: Evans, E., *Gaulish Personal Names*, Oxford, 1967, 472: *LA* I. 14 (1972), 324–27; *Britannia* IX (1978), 383–92.

BROMHAM Brit., *Belgae* (Wiltshire, GB) 51°23′N, 2°02′W
Villa, corridor. Baths, hypocausts, mosaics.
VCH Wiltshire I (1957), 51: Rainey, 1973, 31.

BROONS Lug., *Coriosolites* (Côtes du Nord, F)

48°19′N, 2°16′W
'Kerhalo', villa, building debris. Coins of Gallienus, Postumus, Tetricus.
BSCN XLVII (1909), 3–4.

BROTONNE (Forêt de) Lug., *Lexovii* (Seine-Maritime, F) 49°27′N, 0°43′E
Gallo-Roman foundations at 'Le Lendin (le Câtelier)'. Villa, numerous foundations, steined wells, mosaics of Orpheus; Musée de Rouen.
Fallue, *MSAN* X (1836), 387–90 (*cf* Atlas, 1836, *Ibid* pl V, f 3–4): Grenier, *Manuel* II. 2 (1934), 806.

BRUNOY Lug., *Parisii* (Essonne, F) 48°42′N, 2°30′E
Villa of the early Empire.
Duval, 1961, 78, 230: Roblin, 1971, 33, 175, 328.

BRY-SUR-MARNE Lug., *Parisii* (Val-de-Marne, F) 48°50′N, 2°31′E
BRIVA (bridge over the Marne).
Cemetery. Terra sigillata, coins of Tetricus.
Duval, 1961, 78, 230: Roblin, 1971, 139, 160.

BÚ Lug., *Carnutes* (Eure-et-Loir, F) 48°48′N, 1°30′E
Temple; cella and gallery, pond; pebble mosaics.
Gallia XXVI (1968), 324.

BUFOSSE *cf* Verneuil-en-Halatte

BUIGNY-L'ABBÉ Bel., *Ambiani* (Somme, F) 50°05′N, 1°56′E
Cemetery, small finds.
Agache, fig. 200: Leduque, *Ambianie*, 171 n 827: Agache-Bréart, 1975, 38: Agache, 1978, 272, 273.

*BUIRE-SUR-L'ANCRE Bel., *Ambiani* (Somme, F) 49°58′N, 2°35′E
Cemetery, Gallo-Roman, Merovingian, Carolingian.
BSAP XXXVI (1937), 74–6; Leduque, *Ambianie*, n. 179.

BUISSIÈRE (La) Bel., *Atrebates* (Pas-de-Calais, F) 50°29′N, 2°34′E
Pottery kiln.
BCDMHPC IX.2 (1972–3), 102.

BULLAINVILLE Lug., *Carnutes* (Eure-et-Loir, F) 48°10′N, 1°31′E
Gallo-Roman foundations.
Nouel, 20.

BURLEY Brit., *Belgae* (Hampshire, GB) 50°52′N, 1°39′W
Pottery (New Forest); active 3rd c.
Arch J XXX (1873), 319–24.

BVRRIVM *cf* Usk

BYNE'S FARM Brit., *Regni* (East Sussex, GB) 50°51′N, 0°29′E
Iron workings: furnaces, pottery, terra sigillata 1st to 2nd c.
Straker, 1931, 358: *SxNQ* XIII (1950), 16–18.

CACHAN Lug., *Parisii* (Val-de-Marne, F) 48°47′N, 2°20′E
CATICANTVS (9th c) – deified water.

Between Arceuil and Cachan, aqueduct for Paris, 2nd half 2nd c.
Grenier, *Manuel* IV.1 (1960), 180–191: Duval, 1961, 171–2: Roblin, 1971, 29, 36, 180.

CADBURY CASTLE Brit., *Durotriges* (Somerset, GB) 51°02′N, 2°32′W
Romano-Celtic temple; military buildings, foundations, pottery and many coins discovered during excavations. Situated in hill fort occupied from 4th c BC to 6th c AD; temple in use 3rd and 4th c AD.
Ant XLI (1967), 50–53: Alcock, L., *By South Cadbury is that Camelot,* London, 1972, *passim.*

CAEN Lug., *Viducasses* (Calvados, F) 49°11′N, 0°22′W
Vicus or minor settlement, possibly a port on the Odon.
Coarse pottery and terra sigillata of the periods of Claudius, Domitian and Hadrian; coins of Caesar, Nero, Severus Alexander: neck of a glass vase, bones, cockleshells.
BSAN L (1946–48), 287; LI (1948–51), 351: *Gallia* IX (1951), 83; XXVIII (1970), 269–71.

CAERLEON Brit., *Silures* (Gwent, GB) 51°36′N, 2°57′W
ISCA.
It. A. 484.4, 484.10, 485.8.
Rav. Cos. 106.24.
RIB 316–94.
PNRB 378.
Important port and legionary fortress (II Augusta). Established *c* AD 75, gradually rebuilt in stone starting *c* AD 100. Legion withdrawn *c* AD 293? Houses, shops, amphitheatre; temples of Mithras, Diana and Jupiter Dolichenus assumed from inscriptions.
Nash-Williams, 1969, 29–33: Boon, G. C., *Isca,* Cardiff, 1972: Rainey, 1973, 31.

CAERMEAD *cf* Llantwit Major

CAERPHILLY Brit., *Silures* (Mid Glamorgan, GB) 51°35′N, 3°13′W
Fort, area 1.2 ha. Remains of defences; pottery. Occupied late 1st to mid-2nd c.
Nash-Williams, 1969, 64–65.

CAERWENT Brit., *Silures* (Gwent, GB) 51°37′N, 2°46′W
VENTA SILVRVM.
It. A. 485.9.
Rav. Cos. 106.22 (*Ventaslurum*).
RIB 309–315.
PNRB 493.
Town, capital of the CIVITAS SILVRVM. Established *c* AD 75, declined early 5th c. Forum and basilica, baths, amphitheatre, two Romano-Celtic temples, two invocations each to OCELOS and LENVS; many houses, shops; pottery, coins, numerous small finds; walls enclosing 18 ha.
Arch LVII (1901) – LXII (1911); LXX (1930), 229–88: *BBCS* XV (1952–4), 159–67: Ross, 1967, 191: Rainey, 1973, 32–8: Wacher, 1975, 375–89.

CAESAROMAGVS *cf* Beauvais, Chelmsford

CAGNY Bel., *Ambiani* (Somme, F) 49°52′N, 2°21′E
?Roman camp, probably farm.
Agache, fig. 175–76, 329, 650, 652: *Gallia* XXIII (1965), 300: Agache-Bréart, 1975, 40: Agache, 1978, *passim*.

CAGNY Lug., *Viducasses* (Calvados, F) 49°09′N, 0°15′W
Small villa; small finds, pottery, tiles.
de Caumont, 1831, 230: *BSAN* LII (1948–51), 357.

CAILLY Lug., *Caletes* (Seine-Maritime, F) 49°35′N, 1°15′E
Small settlement. At 'Le Capitole', numerous foundations. Along the road CD 44, cremation cemetery 1st to 3rd c. Burials along the road to Fontaine-le-Bourg.
Cochet, *S.I.*, 195; *BCASM* III (1876), 165.

*CAIX Bel., *Ambiani* (Somme, F) 49°49′N, 2°39′E
Gallo-Roman foundations.
Vasselle, 302: Agache, 1978, 367, 369.

CALA *cf* Chelles

CALAIS Bel., *Morini* (Pas-de-Calais, F) 50°57′N, 1°50′E
Gallo-Roman foundations at the confluence of the rivers Hames and Guines. A tower?, coins. Small finds on the citadel esplanade.
Leduque, *Boulonnais*, 83; *Morinie*, 72: Ringot, 172: Delmaire, 1976, 326 etc.

CALDICOT Brit., *Silures* (Gwent, GB) 51°35′N, 2°46′W
Pottery kiln, late 3rd c.
Mon Ant II (1965), 62–63.

CALETES or CALETI
Caes. *BG* II, 4 (*Caletos*); VII, 75 (*Caletes*); VIII, 7 (*Caletos*).
Str. IV,1,14 (C. 189) (Καλέτους); IV,3,5 (C. 194) (Κάλετοι).
Plin. *NH* IV, 107 (*Galetos*); XIX, 8 (*Caleti*).
Ptol. II,8,5 (Καλῆται).
Oros. VI, 7 (*Caleti*); VI, 11 (*Saletos*).
Not. Tiron. LXXX (*Caletus*).
CIL XIII.1, pp 513–7.
Belgic tribe included in Lugdunensis, occupying the Pays de Caux with capital at IVLIOBONA (Lillebonne, *qv*). Under the late Empire amalgamated with the *Veliocasses* to form the CIVITAS ROTOMAGENSIVM (*cf* Rouen).
Mangard, M., *La tombe gallo-romaine chez les Calètes et les Veliocasses, 1er siècle après J.-C.*, Rouen: Wheeler, R. E. M., and Richardson, K. M., *Hill-Forts of Northern France*, Oxford, 1957.

CALLEVA ATREBATVM *cf* Silchester

CALLOW HILL Brit., *Dobunni* (Oxfordshire, GB) 51°52′N, 1°24′W
Villa, corridor. Enclosing ditch dug late 1st c. Pottery, tiles, coin of Claudius II. Occupied at least from 1st to 3rd c.
Oxon XXII (1957), 11–53.

CAMBLAIN-L'ABBÉ Bel., *Atrebates* (Pas-de-Calais, F) 50°22′N, 2°38′E
Kiln.
Jelski, 144: Agache-Bréart, 1975, 137.

CAMBLIGNEUL Bel., *Atrebates* (Pas-de-Calais, F) 50°23′N, 2°37′E
Kiln.
Jelski, 144.

CAMBRON Bel., *Ambiani* (Somme, F) 50°07′N, 1°46′E
Villa of Marca at 'La Croix qui corne'; columns.
Rancon, Abbé, *BSÉA*, 1909: Vasselle, 302: Agache, fig. 240, 626: Leduque, *Ambianie*, 84, n 452–3: Agache-Bréart, 1975, 40.

CAMERTON Brit., *Belgae* (Avon, GB) 51°18′N, 2°27′W
RIB 180.
Major settlement on road between Bath and Ilchester; remains found over about 12 ha. Workshops for iron-smelting and pewter-making; glass, pottery, coins. Occupation pre-Roman and then from 1st to 5th c, especially late 3rd and 4th c.
Wedlake, W. J., *Excavations at Camerton*, Bath, 1958.

*CAMON Bel., *Ambiani* (Somme, F) 49°53′N, 2°21′E
Gallo-Roman foundations. Shaft of a column, Christian-type inscription VRIS. Coins, statuette, cemetery.
Agache, fig. 284: Leduque, *Ambianie*, n. 302–03: Vasselle, 302: Agache-Bréart, 1975, 40: Agache, 1978, 101.

CAMVLODVNVM *cf* Colchester

CANCALE Lug., *Coriosolites* (Ille-et-Vilaine, F) 48°41′N, 1°51′W
Gallo-Roman foundations. Foot of a large bronze vessel.
Langouet, 1973, 151.

CANETONVM *cf* Berthouville

CANEWDON Brit., *Trinovantes* (Essex, GB) 51°36′N, 0°43′E
Salt workings, briquetage.
Nenquin, 1961, 79: *VCH Essex* III (1963), 61.

CANNES-ÉCLUSE Lug., *Senones* (Seine-et-Marne, F) 48°22′N, 2°59′E
Milestone on the Sens-Paris road; Wuilleumier, 482.
Large villa at 'Les Bagneux'; 1st c to about 250, the second quarter of 4th c, near a large protohistoric and Gallo-Roman cemetery and a Bronze Age settlement.
Gallia XXI (1963), 362; XXIII (1965), 316.

CANONIVM *cf* Kelvedon

CANOUVILLE Lug., *Caletes* (Seine-Maritime, F) 49°48′N, 0°36′E
Small settlement to the north of the road CD 71: cremation cemetery, theatre.
BSNEP XXIV (1919–21), 53–6, *cf* 22, 52: *Gallia* XXXVII (1979), 237–46.

CANTELEU Lug., *Veliocasses* (Seine-Maritime, F) 49°28′N, 1°00′E
Temple, in the forest of Roumare, canton 'Hazard'. Also 'villa' and Frankish tombs.
BCASM XI, 204: Vesly, 14.

CANTERBURY Brit., *Cantiaci* (Kent, GB) 51°16′N, 1°04′E
DVROVERNVM CANTIACORVM.
Ptol. II,3,12 (Δαρούερνον).
It. A. 472.5 (*Duroruerno*), 473.4 (*Durarueno*), 473.9 (*Durarueno*).
Tab. Peut. (*Duroaverus*).
Rav. Cos. 106.36 (*Duroaverno Cantiacorum*).
RIB 15.
PNRB 353–4.
Important town, capital of the CIVITAS

CANTIACORVM, following a Claudian fort which succeeded a pre-Roman settlement. Walls enclosing c 52 ha; potters' workshops, baths, tile kilns and limekilns, cemetery, theatre. Port at Fordwich, qv. Museum.
VCH Kent III (1932), 61–80: JRS XXXVIII (1948) et seq: Ant XXIII (1949), 153: Ant J XXXVI (1956), 40–56: Ross, 1967, 49: Britannia I (1970), 73, 83–113, 183: Wacher, 1975, 178–95; T. Tatton-Brown, S. S. Frere et al., The Archaeology of Canterbury I, II, Canterbury, 1982.

CANTIACI
Caes. BG V, 13, 14, 22 (Cantium).
Str. I,4,3 (C. 63); IV,3,3 (C. 193); V,1 (C. 199) (Κάντιον).
Ptol. II,3,12 and 13 (Κάντιοι).
Rav. Cos. 106.36 (Duroaverno Cantiacorum).
RIB 192 (ci(vis) Cant . . .).
PNRB 299–300.
Cantium was a geographical rather than a political expression; thus Caesar states that there were four kings in Kent, and refers to its inhabitants as qui Cantium incolunt, not as Cantii or Cantiaci. The CIVITAS CANTIACORVM was probably an artificial creation of the Romans; it comprised all Kent, with parts of Surrey and Sussex, and had its capital at DVROVERNVM (Canterbury, qv).
Rivet, 1964, 144–5: Frere, 1978, 234–5.

CANTIVM PROMONTORIVM cf South Foreland

CANVEY ISLAND Brit., Trinovantes (Essex, GB) 51°30′N, 0°36′E
Salt working; briquetage, pottery 1st to 4th c.
TEAS II (1966), 14–33, 158, 329: Nenquin, 1961, 80: VCH Essex III (1963), 62.

CANVILLE-LES-DEUX-ÉGLISES Lug., Caletes (Seine-Maritime, F) 49°46′N, 0°51′E
Cemetery at 'La Garenne'.
Cochet, S.I., 431.

CANY-BARVILLE Lug., Caletes (Seine-Maritime, F) 49°47′N, 0°38′E
Minor settlement, cross-roads; important foundations (a quadrangular Roman building with thick walls at the centre of Cany-Barville in an island near the Pont des Moulins. Cremation cemetery on the Vittefleur road. A villa at the hamlet of Vinfrainville.
Cochet, S.I., 448 sq.; Rép., col. 478: Deglatigny, 1931, 28.

CARACOTINVM cf Harfleur
CARANTOMAGVS cf Charenton
CARDIFF Brit., Silures (South Glamorgan, GB) 51°29′N, 3°11′W
Fort and minor settlement. The fort, at Cardiff Castle, was probably occupied under the Flavians and again from c AD 300 to 367. Pottery and a coin of Trajan found in the settlement.
Nash-Williams, 1969, 70–73.

CARENTAN Lug., Unelli (Manche, F) 49°18′N, 1°15′W
CROCIATONVM.
Ptol. II,8,2 (Κρουκιάτοννον).
Tab. Peut. (Crouciaconnum).
Oderic Vital. Hist., XX (Vicus Carentonus).
'le Mur du Quai', 'Pont St Hilaire', Vicus or small settlement with port, on the road between Bayeux and Valognes; flourished 2nd and 3rd c. Tiles, bricks, bronze hatchet, buckles, coins, silver coin of Philip; dug-out canoe.
de Gerville, 1854, 96: BSAIC XXIV (1900), 137–8: BSNÉP XIII (1905), 152–3.

CARHAIX Lug., Osismii (Finistère, F) 48°17′N, 3°35′W
VORGIVM, later OSISMIS.
?Ptol. II,8,5 (Οὐοργάνιον, Οὐοργόνιον, Οὐόργον).
Tab. Peut. (Vorgium).
CIL XIII 9013 (milestone of Mael Carhaix).
cf OSISMII.
Town, capital of the CIVITAS OSISMIORVM. Junction of many roads. Various foundations throughout the town; aqueduct 30 km long; burials. Abundant imported pottery, Arretine, Argonne 4th c; a gold ring, many silver paterae; numerous coins from the end of the Republic to the 4th c.
BSAF, 1875, 124: du Chatellier, 1907, 162–3: Ann Bret LXII (1955), 181–201; LXVIII (1971), 165–87; LXIX (1972), 149–59: Pape, 1969, 10–12.

CARISBROOKE CASTLE Brit., Belgae (Isle of Wight, GB) 50°41′N, 1°19′W
Possible Saxon shore fort under medieval castle. Coin of Constantine.
JRS XVI (1926), 235: Johnson, 1979, 143.

CARISBROOKE VICARAGE Brit., Belgae (Isle of Wight, GB) 50°41′N, 1°19′W
Villa; baths, hypocausts, mosaics.
VCH Hampshire I (1900), 316–17: Britannia I (1970), 300: Rainey, 1973, 39.

CARMARTHEN Brit., Demetae (Dyfed, GB) 51°52′N, 4°18′W
MORIDVNVM.
Ptol. II,3,12 (Μαρίδουνον).
It. A. 482.9 (Muridono), 483.5 (Muriduno).
RIB 412–3.
PNRB 422.
Fort and town, the latter probably capital of the CIVITAS DEMETARVM. Fort occupied under the Flavians, town from late 1st to 4th c. Walls, enclosing c 6 ha.; buildings, amphitheatre.
Ross, 1967, 88: Nash-Williams, 1969, 23–6: Carm Ant V (1969), 2–5; VI (1970), 4–14: Wacher, 1975, 389–393: Britannia VIII (1977), 360; IX (1978), 408; X (1979), 272–3.

CARNANTON Brit., Dumnonii (Cornwall, GB) 50°26′N, 4°59′W
Tin mine; inscribed ingot found.
Cornish Archaeology VI (1967), 29–31.

CARNVTES
Caes. BG II, 35; V, 25, 29, 56; VI, 2, 3, 4, 13, 44; VII, 2, 3, 11, 75; VIII, 4, 5, 31, 38, 46.
Str. IV,2,3 (C. 191); IV,3,4 (C. 193) (τῶν Καρνούτων).
Liv. V, 34.
Tibullus I,7,12 (Carnuti).
Plin. NH IV, 107 (Carnuteni foederati).
Plut. Caesar, 25 (Καρνουτῖνοι).
Flor. I,45(III,10),20.
Ptol. II,8,10 (Καρνοῦνται).
Not. Dig. Occ. XLII, 33 (praefectus laetorum Teutonicianorum, Carnunta).
Not. Gall. IV, 3 (civitas Carnotum).
Sulp. Sev. Dial II,2,3 (in Carnutena civitate); II,4,5 (Carnotum oppidum).
Oros. VI, 11.
Conc. Aurel. a. 511 (episcopus ecclesiae Carnotenae).
Ven. Fort. Carmina IV, 7; Vita S Martini III, 153.

Greg. Tur. *HF* IV, 34 (49); V, 25 (34); VII, 2, 17; VIII, 10; IX, 5, 20.
CIL XI, 716 (*Carnutino*); XIII.1, 1672 (*Carn . . ., bis*); 1694 (*Carnut(i)*); 2010 (*ci(vi) Carnutino*); ?2011 (*civ(is) Ca[rnutus?]*); notes, p 492.

Large tribe of Celtic Gaul, included in Lugdunensis (later Lugdunensis Senonia), with capital at AVTRICVM, later known as CIVITAS CARNOTVM (Chartres, *qv*). The town of CENABVM (Orléans, not on this map) was also in their territory under the early Empire, but later became a *civitas*-capital in its own right.

Braemer, F., Les relations commerciales des C. d'après les découvertes monétaires, *Actes du Congrès international de numismatique*, Paris, 1953 (1957), II, 563–72.

CASSEL Bel., *Menapii* (Nord, F) 50°48′N, 2°29′E
CASTELLVM MENAPIORVM.
Ptol. II,9,5 (Κάστελλον).
It. A. 376.5, 377.2, 377.6 (*Castello*).
Tab. Peut. (*Castello Menapiorii*).
CIL XIII, 9158 (itinerary of Tongres).

Town, capital under the early Empire of the CIVITAS MENAPIORVM; *burgus* of the late Empire. Roads to Tournai, Boulogne and Cologne. Rampart, 3rd c; important cemetery on the south flank; statue of Galba.

BCHN XXVI (1904), 217–22: Blanchet, *Enceintes*, 125: Ringot, 170: Leduque, *Morinie*, 82.

*CASTEL Bel., *Ambiani* (Somme, F) 49°46′N, 2°27′E
Cemetery.
Leduque, *Ambianie*, 145, n 351.

CASTELLVM MENAPIORVM *cf* Cassel

CASTLE FIELD *cf* Worth

CASTLEFIELD Brit., *Belgae* (Hampshire, GB) 51°13′N, 1°26′W
Villa, aisled.
VCH Hampshire I (1900), 302.

CATENOY Bel., *Bellovaci* (Oise, F) 49°23′N, 2°30′E
Roman fort (Caesar) at oppidum of the Bellovaci (2nd campaign against the Bellovaci).
BSPF XIV (1917), list LXI, 469: *DAG*.

CATVLLIACVS *cf* St Denis

CATVVELLAVNI
Ptol. II,3,11 (Κατυευχλανοί).
Cass. Dio LX,20,2 (Κατουελλανοί).
RIB 1065 (*natione Catvallauna*); 1962 (*civitate Catuvellaunorum*).
PNRB 304–5.

British tribe, which from an original nucleus in Hertfordshire came to dominate most of south-eastern Britain in the pre-conquest period. Recognised by the Romans as a *civitas*, with territory reaching from the Thames to Leicestershire and capital at VERVLAMIVM (near St Albans, *qv*).

Rivet, 1964, 145–8; Frere, 1978, 57–65 *passim*; Webster, 1980, *passim*; Salway, 1981, 43, 47, 55–9, etc.

CAUDEBEC-EN-CAUX Lug., *Caletes* (Seine-Maritime, F) 49°43′N, 0°44′E
?LOTVM
It. A. 382.2.

Minor settlement. Cremation cemetery at 'Côte Saint-Clair'; ancient debris at 'Côte de la Vignette'; at Mont Calidu, pre-Roman oppidum with traces of Gallo-Roman occupation.

Cochet, *S.I.*, 477: *BCASM* III, 356; V, 452, 484: *DAG*; *BSPF* XVI (1919), list lxxvii, 186.

CAUDEBEC-LÈS-ELBEUF Lug., *Aulerci Eburovices* (Seine-Maritime, F) 49°18′N, 1°01′E
VGGADE
It. A. 384.3.

Major settlement, important in 1st c. declined AD 275. Numerous traces of buildings (including a temple); cremation cemetery (1st-3rd c) at 'Le bout de la Ville', with inhumation cemetery superimposed; cremation at 'côte Piéton'; rubbish dump at 'la Fosse à Moules'; bath-house at 'la Mare-aux-Boeufs'. Coin hoard, 2nd half of 3rd c.

Saint-Denis, M., *Histoire de C.-l.-E.*: Cochet, *Sépultures gauloises*, 110: *BSNÉP* XX (1912), 100–05; XXXIX.2 (1968), 91: Mathière, 1925, 340–48: *Ann Norm* XV (1965), no 3, 437: *Gallia* XXIV (1966), 266; XXVI (1968), 369, XXX (1972), 345.

CAULNES Lug., *Coriosolites* (Côtes du Nord, F) 48°17′N, 2°09′W
'Gare de Caulnes', villa; baths, fibulae. Coins from Antoninus Pius to Valens; statuettes in white clay.
RA IX (1864), 414–9: *BSCN* III (1865), 31–8; XLVII (1909), 9–11.

CELLOVILLE Lug., *Veliocasses* (Seine-Maritime, F) 49°23′N, 1°11′E
Villa, 'Le Thuit' (SERBONIS VILLA, in property register). Hypocaust, heat-ducts, pavements, terra sigillata.
Vesly, *BSNÉP* XII (1904), 186; XVII (1909), 32 sq: *BCASI* XIII (1905), 261 sq; XV (1909), 112.

CENOMANI *cf Aulerci*

CERISY-BULEUX Bel., *Ambiani* (Somme, F) 49°59′N, 1°44′E
Temple.
Agache-Bréart, 1975, 43.

CERLANGUE Lug., *Caletes* (Seine-Maritime, F) 49°29′N, 0°26′E
Cremation cemetery at the hamlet of Claque.
Cochet, *S.I.*, 388.

CHADWELL ST MARY Brit., *Trinovantes* (Essex, GB) 51°28′N, 0°22′E
Pottery kiln.
VCH Essex III (1963), 63.

CHAINTREAUX Lug., *Senones* (Seine-et-Marne, F) 48°12′N, 2°49′E
Gallo-Roman foundations, two dwellings, at 'bois des Pitelliers'.
Nouel, 19, 26.

CHALK Brit., *Cantiaci* (Kent, GB) 51°25′N, 0°26′E
Several pottery kilns, 1st and 2nd c; villa.
Arch Cant LXXIII (1959), 220–23: *JBAA* IV (1849), 393–94: *VCH Kent* III (1932), 130–1; *Britannia* III, (1972), 112–48.

CHAMANT Bel., *Silvanectes* (Oise, F) 49°13′N, 2°37′E
Gallo-Roman foundations in an enclosure; another enclosure at 'Mont Alta'. In the vicinity of the Temple de la forêt d'Halatte, *qv*.

CHAMPS GERAUX (Les) Lug., *Coriosolites* (Côtes du Nord, F) 48°25′N, 1°58′W
'le Grand Bois', Gallo-Roman foundations, debris.
Gallia XXXIII (1975).

CHANCTONBURY RING Brit., *Regni* (West Sussex, GB) 50°54′N, 0°23′W
Romano-Celtic temple, in Iron Age hill fort. Coins of 1st, late 3rd and 4th c.
SxAC LIII (1909), 131–37: *VCH Sussex* III (1935), 52–53: Lewis, 1966, *passim*: *Britannia* XI (1980), 173–222.

CHAPELLE CHAUSSÉE (La) Lug., *Redones* (Ille-et-Vilaine, F) 48°16′N, 1°51′W
'la Plesse', Gallo-Roman foundations.
Bull Arch Ass Bret, 1887, 229.

CHAPELLE-ST-FRAY (La) Lug., *Aulerci Cenomani* (Sarthe, F) 48°07′N, 0°05′E
Villa near 'Le Vau'.
Bouton.

CHARENTON-LE-PONT Lug., *Parisii* (Val-de-Marne, F) 48°49′N, 2°25′E
CARANTOMAGVS, PONS CARANTONIS (6th c.).
Gallo-Roman foundations.
Duval, 1961, 78, 88, 90, 203: Roblin, 1971, 35, 109, 161, 215.

CHARLEVAL *cf* ?*Ritumagus*

CHARTERHOUSE Brit., *Belgae* (Somerset, GB) 51°18′N, 2°43′W
?ISCALIS
Ptol. II,3,13 (Ἰσχαλις).
RIB 184–6.
PNRB 379.
Town (with early Roman fort, based on lead-mining. Amphitheatre, buildings; pottery, coins, glass, many small finds. Occupied throughout the Romano-British period and probably before.
VCH Somerset I (1906), 334–44: *Britannia* II (1971), 277–8: Branigan-Fowler, 1976, 184–5.

CHARTHAM HATCH Brit., *Cantiaci* (Kent, GB) 51°16′N, 0°59′E
Iron workings; furnaces, pottery 1st c.
Jessup, 1970, 189: *KAR* XIX (1970), 11.

CHARTRES Lug., *Carnutes* (Eure-et-Loir, F) 48°27′N, 1°30′E
AVTRICVM (from AVTVRA FLVMEN, river Eure, *qv*), later CIVITAS CARNOTVM, VRBS CARNOTENA.
Ptol. II,8,10 (Αὔτρικον).
Tab. Peut.
CIL VIII, 1876; XIII.1, p 473.
Cf CARNVTES.
Town, capital of the CIVITAS CARNVTVM. Amphitheatre 2nd c, hypocausts, market with small shops, or forum?; aqueducts, cemetery, statuary, workshop of maker of sculptured figurines; more than 100 votive pits in the Place de la République. Museum.
Ver-lès-Chartres: 'Houdouenne', aqueduct of Chartres: 'Loché', pool, canals, mosaics, coins, bust of Vesta.

Blanchet, *Enceintes*, 67–8: Duval, 1961, 91, 117, 176, 194, 243, 245, 285: Nouel, 16, 17, 26, 27, 29, 30: *BSAEL* XXXIX.4 (1970); LXIX.4 (1977): *Gallia* XXI (1963), 395; XXIV (1966), 239; XXVI (1968), 321; XXVIII (1970), 253; XXX (1972), 312.

CHÂTEAUDUN Lug., *Carnutes* (Eure-et-Loir, F) 48°05′N, 1°20′E
DVNVM (6th c.).
Gallo-Roman foundations. Mosaic in the municipal museum.
Duval, 1961, 96: Nouel, 7–10, 21, 26.

CHÂTEAU-LANDON Lug., *Senones* (Seine-et-Marne, F) 48°09′N, 2°42′E
?VELLAVNODVNVM (*qv*).
Gallo-Roman foundations at a pre-Roman hill-fort.
Jalmain, 90: Nouel, 21, 22, 44.

CHÂTEAUNEUF DU FAOU Lug., *Osismii* (Finistère, F) 48°11′N, 3°49′W
'Cizty' = 'la vieille maison', villa near the road between Carhaix and the neighbourhood of Crozon. Numerous sestertii; an aureus of Antoninus Pius (158). 2nd-3rd c.
BSAF, 1971, 26.

CHÂTRES-EN-BRIE Lug., *Parisii* (Seine-et-Marne, F) 48°42′N, 2°49′E
Roman fort, ?frontier post of the late Empire.
Duval, 1961, 223: Roblin, 1971, 15.

***CHAULNES** Bel., *Ambiani* (Somme, F) 49°49′N, 2°48′E
Temple; cellar or funerary vault. Foundations of a destroyed structure; in the interior, roof debris, tiles, vessels.
BSAP XVI (1887), 375: Leduque, *Ambianie*, 31, n.109.

CHAUSSÉE-TIRANCOURT (La) Bel., *Ambiani* (Somme, F) 49°57′N, 2°09′E
Gallo-Roman foundations. 'Le Grand Fort', 'Camp de César', 'Fossé Sarrazin', oppidum of the Ambiani at the confluence of the Somme and the Acon. Outside, Roman trenches and works. Sarcophagi, coins, pottery.
DAG: *BSPF* XVII (1920), list LXXIX, 56: *RN* XLIV (1962), 323: Agache, fig. 6, 7, 40, 86, 94, 372, 373: *Celticum* XV (1966), 35–46: Leduque, *Ambianie*, 74, n. 334, 340, 342–6: Agache-Bréart, 1975, 43–5: Agache, 1978, *passim*.

CHEDWORTH Brit., *Dobunni* (Gloucestershire, GB) 51°49′N, 1°55′W
RIB 126–8.
Villa and probable cult centre: two temples, one Romano-Celtic, one described as a nymphaeum; dedications to OLLVDIVS and LENVS MARS. Two bath suites, many mosaics, pottery, reliefs, bronzes, many small finds. Occupied early 2nd c to *c* AD 500. Museum.
Arch J XLIV (1887), 322–36: *TBGAS* LII (1930), 255–64; LXXVIII (1959), 5–23: Ross, 1967, 50 etc.: R. Goodburn, *The Roman Villa, Chedworth*, 1972: Rainey, 1973, 40: McWhirr, 1981, 90–92, 150–53.

CHELLES Lug., *Parisii* (Seine-et-Marne, F)

48°53′N, 2°36′E
CALA.
Greg. Tur. *Hist. Francorum* V,29; VI,33; VII,4; X,19.

Large villa. Occupation of the site from pre-Roman to Merovingian times. Gallo-Roman foundations on the site of the abbey; rubbish dumps; brick walls, pottery, tiles, Gaulish and Roman coins. Alfred Bonno museum.

Duval, 1961, 78, 230: *Gallia* XXV (1967), 215; XXVIII (1970), 243; XXX (1972), 304.

CHELMSFORD Brit., *Trinovantes* (Essex, GB) 51°43′N, 0°28′E
CAESAROMAGVS.
It. A. 474.3 (*Caesaromago*); 480.6 (*Cesaromago*).
Tab. Peut. (*-baromaci*).
Rav. Cos. 106.51 (*Cesaromago*).
PNRB 287.

Town on the road between London and Colchester, possibly capital of the CIVITAS TRINOVANTVM (but *cf* Colchester). Location of a fort of Aulus Plautius. Destroyed by Boudica AD 60; reconstructed. Bath-houses, houses, mosaics, cemetery, Romano-Celtic temple. The Latin name is of a type (imperial title + Celtic suffix) otherwise unknown in Britain but widespread in Gaul (*cf* Bayeux, Beauvais, Lillebonne).

EHR LII (1937), 198: *VCH Essex* III (1963), 5, 63: Rivet, 1964, 147: *JRS* LIX (1969), 223: *Britannia* I (1970), 183: *TEAS* II (1970), 333: Wacher, 1975, 195–202: Dunnett, 1975, 81–6: Rodwell-Rowley, 1975, 159–73.

CHEMAULT Lug., *Carnutes*? (Loiret, F) 48°03′N, 2°21′E
Gallo-Roman foundations of burnt buildings at 'Champ carré'; human and animal bones, mosaics, a capital (in the museum at Pithiviers), quern.
Nouel, 17, 29, 31.

CHERBOURG Lug., *Unelli* (Manche, F) 49°38′N, 1°38′W
CORIALLVM or CORIALLO.
Tab. Peut. (*Coriallum* or *Coriallo*).
Chron. Fontenelle (*pagus Coriovallensis*).

'Guévillon', 'Montagne du Roule', 'les Mielles de Tourlaville'. Port, 1st to 5th c. Two coin hoards of early and late Empire. Wells, burials; a gold belt; coins from Tiberius to Constantine II, and many thousands of coins in the sands. Terra sigillata, NAMILIANI, SATTO F (*CIL* XIII, 10010/1405, 1734); pottery drainpipes, figure, weights.

At Tourlaville, a temple. Foundations, wells, metal objects, terra sigillata, 15 terracotta statuettes of Venus, six figurines of horses, seated mother-goddess, head of a laughing infant, head of Apollo; iron objects, querns, bronze rider-god, bronze Ceres, bronze Mercury; 400 coins, Augustus to Constantine; coin hoard weighing 30 kg, up to Justinian II.

Asselin, 1829, 1832, *passim*: *MSAN* V (1829–30), lxiii-lxv; VI (1831–2), 450–52: de Gerville, 1854, 101: Renault, 1880, 21: Blanchet, 1900, 206 nn 423–4: *BSAIC* XXIV (1900), 22–36: *BSNÉP* XIII (1905), 137–43: *BSAN* LIII (1955–6), 197–263: Bouhier, 1962, *v* Cherbourg.

CHESTERS (The), Gloucestershire *cf* Woolaston

CHEVILLY-LARUE *cf* Haÿ-les-Roses (L')

CHEVRAINVILLIERS Lug., *Senones* (Seine-et-Marne, F) 48°15′N, 2°37′E
Gallo-Roman foundations, tiles, querns, terra sigillata.
Nouel, 21, 31.

CHEW STOKE *cf* Pagans Hill

CHICHESTER Brit., *Regni* (West Sussex, GB) 50°50′N, 0°47′W
NOVIOMAGVS REGNORVM (or REGNENSIVM).
Ptol. I,15,7; II,3,13 (Νοιόμαγος).
It. A. 477.10 (*Regno*).
Rav. Cos. ?106.17 (*Noviomagno*); 106.20 (*Navimago Regentium*).
RIB 89–96.
PNRB 427.

Town and port, capital of the CIVITAS REGNORVM or REGNENSIVM. Roads to London, Silchester and Bitterne. Occupied 1st to 5th c. Military equipment of 1st c; temple of Neptune and Minerva (*RIB* 91); amphitheatre; Jupiter column; walls enclosing *c* 40 ha; cemeteries; many small finds.

Ross, 1967, 196: Down, A., and Rule, M., *Chichester Excavations* I, Chichester, 1971; Down, A., *ibid.* II, 1974; III, 1978; V, 1981: Rainey, 1973, 43: Cunliffe, 1973, 47–68: Wacher, 1975, 239–55.

CHIGNALL ST JAMES Brit., *Trinovantes* (Essex, GB) 51°46′N, 0°25′E
Villa, courtyard, identified from air.
Britannia VI (1975), 263–4; VIII (1977), 405–7; IX (1978), 449; X (1979), 308; XI (1980), 376; XII (1981), 348; XIII (1982), 371.

CHIGWELL Brit., *Trinovantes* (Essex, GB) 51°37′N, 0°07′E
?DVROLITVM (but *cf* Romford).
It. A. 480.7. (*Durolito*).
PNRB 352.

Major settlement on the road between London and Great Dunmow. Foundations, mosaics, cemetery.
VCH Essex III (1963), 88: Rivet, 1964, 147: *Britannia* VI (1975), 81, 93; Rodwell-Rowley, 1975, 85–101 *passim*.

CHILGROVE Brit., *Regni* (West Sussex, GB) 50°54′N, 0°49′W
'Brick Kiln Farm' Villa; corridor; baths, mosaics; pottery and coins early 2nd to late 4th c. Iron Age huts on same site.
50°55′N, 0°48′W
'Cross Roads Field' Villa; corridor; aisled, detached bath-building; mosaics. Reconstructed late 3rd c.
Britannia II (1971), 286: Rainey, 1973, 43–4: Cunliffe, 1973, 87–8. A. Down, *Chichester Excavations* IV, 1979.

*CHILLY Bel., *Ambiani* (Somme, F) 49°47′N, 2°46′E

Small settlement, with temple, 'Le Bois du Carme, Canton de Justice'.
Agache, 1972, 321; 1978, 44, 378, 398, 402, 422.

CHILMARK Brit., *Belgae* (Wiltshire, GB) 51°05′N, 2°03′W
Stone quarry. Close by to west a votive deposit of 4,674 sherds, iron nails and other objects, amber bead, shale bracelet, parts of querns, 2 coins – Urbs Roma and Claudius Gothicus or Tetricus.
VCH Wiltshire I (1957), 56.

***CHIRMONT** Bel., *Ambiani* (Somme, F) 49°42′N, 2°24′E
Gallo-Roman foundations, pottery, at 'Le Vieux Chirmont'.
Vasselle, 306: Agache-Bréart, 1975, 45: Agache, 1978, 113, 272, 275, 282.

CHITCOMBE Brit., *Regni* (East Sussex, GB) 50°57′N, 0°34′E
Iron workings; furnaces, pottery, tiles, terra sigillata.
Straker, 1931, 345–47: *SxNQ* VI (1937), 205.

CHOISY-LE-ROI Lug., *Parisii* (Val-de-Marne, F) 48°46′N, 2°25′E
Cemetery, 3rd c; glass urns, sarcophagi, wooden figures, bases of Gallo-Roman vessels.
Duval, 1961, 232: Roblin, 1971, 58, 188, 267.

CICVCIVM *cf* ?Y Gaer

CIRENCESTER Brit., *Dobunni* (Gloucestershire, GB) 51°43′N, 1°58′W
CORINIVM DOBVNNORVM.
Ptol. II,3,12 (Κορίνιον or Κορίννιον).
Rav. Cos. 106.31 (*Cironium Dobunorum*).
RIB 101–18.
PNRB 321.
Town; capital of the CIVITAS DOBVNNORVM, on the river Churn, with roads to Bath and Leicester (Fosse Way), Gloucester, Silchester and St Albans (Akeman Street). Military occupation *c* AD 45 to 75? (*RIB* 108, 109). Town developed under the Flavians eventually second largest town in Britain; in the 4th c probably capital of BRITANNIA PRIMA (*RIB* 103); survived into 5th c, taken by the Saxons after the Battle of Dyrham AD 577. Centre of the 'Corinian School' of mosaicists. Walls enclose area of 96 ha. Forum, basilica, market, shops, houses, amphitheatre, possible theatre, cemeteries. Sculptures and reliefs including Jupiter column; numerous finds.
Ross, 1967, 36 etc.: *Ant J* XLI (1961) to XLVII (1967) inclusive, XLIX (1969) and LIII (1973): Rainey, 1973, 21, 44–53: Wacher, 1975, 289–315: Branigan-Fowler, 1976, 81–98: McWhirr, A., *Roman Gloucestershire*, Gloucester, 1981, 21–58: Wacher, J., and McWhirr, A., (ed), *Cirencester Excavations I*, 1981. II, 1982.

***CLAIRY-SAULCHOIX** Bel., *Ambiani* (Somme, F) 49°51′N, 2°10′E
Cemetery.
Agache, fig. 203, 416: Leduque, *Ambianie*, 156, n.550: Agache-Bréart, 1975, 46–47: Agache, 1978, 167, 261, 272, 316, 330.

CLANVILLE Brit., *Belgae* (Hampshire, GB) 51°14′N, 1°33′W
RIB 98.
Villa; courtyard; mosaics, window glass, painted plaster, coins of late 3rd and early 4th c. Possible milestone of Carinus.
VCH Hampshire I (1900), 296: Rainey, 1973, 53: Sedgley, 1975, 19.

CLATWORTHY *cf* Syndercombe

CLAVSENTVM *cf* ?Bitterne

CLEAR CUPBOARD *cf* Farmington

CLEARWELL Brit., *Dobunni* (Gloucestershire, GB) 51°46′N, 2°37′W
Iron workings; coins mostly 3rd c.
TBGAS XXIX (1906), 12.

CLEAVEL POINT Brit., *Durotriges* (Dorset, GB) 50°40′N, 2°00′W
Pottery kilns; 1st and 2nd c.
RCHM Dorset II.3 (1970), 597–98: *Britannia* XI (1980), 390.

CLEDEN CAP SIZUN Lug., *Osismii* (Finistère, F) 48°03′N, 4°39′W
'Trouguer', Iron Age promontory fort, near two roads leading from Cap Sizun to Quimper and Douarnenez respectively. Foundations; many small finds including a bronze statuette; many coins of 2nd and 3rd c. Coarse pottery and terra sigillata.
BSAF, 1875, 124: *Gallia* XII (1954), 156–60; XIII (1955), 153–55; XV (1957), 181–85: Wheeler, R. E. M., and Richardson, K. M., *Hill-Forts of Northern France*, Oxford, 1957, 109.

CLEGUEREC Lug., *Veneti* (Morbihan, F) 48°07′N, 3°05′W
At 'Kerfulus', Gallo-Roman foundations, tegulae, potsherds. At 'Locmaria', brick coffins 2.20 m long containing skeletons. Tegulae, 4th-c potsherds.
BSPM, 1901, 293–95.

CLERMONT-DE-L'OISE Bel., *Bellovaci* (Oise, F) 49°23′N, 2°24′E
Roman fort claimed at 'Bois des Côtes de Nointel', dating from Caesar's campaign against the *Bellovaci*, 51 BC, disproved 1959.
RÉA XXXIX (1937), 347–62; XLII (1940), 645–52: *Mém Soc Clermont*, 1939: *Gallia* I (1943), 81–127: *BSNAF*, 1959, 263.

CLIFFE Brit., *Cantiaci* (Kent, GB) 51°28′N, 0°30′E
Two pottery kilns, 3rd c.
Arch Cant LXXXI (1966), liv.

CLINCHAMPS Lug., *Viducasses* (Calvados, F) 49°05′N, 0°24′W
Villa on the road between Vieux and Seés: 2nd c.
de Caumont, 1831, 227; 1838, 159–62; 1850, 176–78.

CLOYES-SUR-LE-LOIR Lug., *Carnutes* (Eure-et-Loir, F) 48°00′N, 1°14′E
Gallo-Roman foundations: querns, lead water-pipes. Cemetery.
Nouel, 20, 27, 31, 33.

COATASCORN Lug., *Osismii* (Côtes du Nord,

F) 48°40′N, 3°15′W

'Coetbriand', Gallo-Roman foundations, débris, potsherds.

MSCN I (1883–84), 342.

COBHAM PARK Brit., *Cantiaci* (Kent, GB) 51°24′N, 0°26′E

Corridor villa; baths, remains of walls, *tesserae*, frescoes.

Arch Cant LXXVI (1961), 88–109.

COCHEREL *cf* Houlbec-Cocherel

COELBREN Brit., *Silures* (West Glamorgan, GB) 51°47′N, 3°39′W

Temporary camp.

JRS LIX (1969), 200; RCAHM (Wales), *Glamorgan* I, ii, 1976, 99–100.

COELBREN GAER Brit., *Silures* (West Glamorgan, GB) 51°47′N, 3°39′W

Fort; occupied from *c* AD 70 to 150.

Nash-Williams, 1969, 81–83, RCAHM (Wales), *Glamorgan* I, ii, 1976, 83–4.

COETMIEUX Lug., *Coriosolites* (Côtes du Nord, F) 48°30′N, 2°36′W

At 'Clos-Rougers', villa; bronze coins of the late Empire: at 'la Tour' and at 'les Airs', Gallo-Roman foundations.

MSCN I (1883–84), 224.

COIGNEUX Bel., *Ambiani* (Somme, F) 50°07′N, 2°33′E

Cemetery; tombs, urns, arms, coins, statuette of Mercury.

Leduque, *Ambianie,* 30, n.78.

COINCES Lug., *Carnutes* (Loiret, F) 48°01′N, 1°45′E

Gallo-Roman foundations.

Nouel, 20, 22.

COLCHESTER Brit., *Trinovantes* (Essex, GB) 51°53′N, 0°54′E

CAMVLODVNVM COLONIA.

Coins of Cunobelinus (Mack, 186, 201–60).

Plin. *NH* II, 187 (*Camaloduno*).

Tac. *Ann.* XII, 32; XIV, 31–2.

Ptol. II,3,11 (Καμουδόλανον).

Cass. Dio LX,21,4 (Καμουλόδουνον).

It. A. 474.4 (*Colonia*); 480.4 (*Camoloduno*).

Tab. Peut. (*Camuloduno*).

Rav. Cos. 106.52 (*Manulodulo colonia*).

CIL III, 11233 (*Camuloduni*); XIV, 3955 (*Colonia Victricensis quae est in Brittannia Camaloduni*).

RIB 63–9

PNRB 294–5, 312.

Important town. Pre-Roman capital of the *Trinovantes,* then of the *Catuvellauni* under their king Cunobelinus. Taken by Claudius AD 43; fort, legionary fortress (Legio XX Valeria) then *colonia* AD 49. Destroyed by Boudica AD 60; rebuilt and given title *Victricensis.* Possible capital of the CIVITAS TRINOVANTVM (but *cf* Chelmsford). Temple of Claudius (Seneca, *Apocol.,* 8) on site now occupied by castle and museum. Seven Romano-Celtic temples, pottery (incl. *terra sigillata*) and tile kilns, dwellings, mosaics, cemetery, *vicus.* Walls enclosing *c* 40 ha; well-preserved Balkerne Gate at the start of the road to London. Temple and theatre at Gosbeck's Farm, *qv*; port at Fingringhoe Wick, *qv*.

Lambrechts, 1942, 129: Hawkes-Hull, 1947, *passim*: Hull, 1958 and 1963, *passim*: Lewis, 1966, *passim*: Ross, 1967, 268, 362–7: *TEAS* II (1967), 37–42: *Arch Camb* XCVII (1968), 77: Dunnett, 1975, 63–81: Wacher, 1975, 104–20: *Britannia* VIII (1977), 65–105: Frere, 1978, *passim*: Webster, 1980, *passim*.

COLD KITCHEN HILL Brit., *Belgae* (Wiltshire, GB) 51°09′N, 2°14′W

Temple, plan unknown; many bronzes and other small finds, coins, chiefly 3rd and 4th c.

VCH Wiltshire I (1957), 48–49: Ross, 1967, 44 etc.

COLERNE Brit., *Belgae* (Wiltshire, GB) 51°27′N, 2°16′W

Villa; mosaics.

VCH Wiltshire I (1957), 59: Rainey, 1973, 56.

COLLEVILLE Lug., *Caletes* (Seine-Maritime, F) 49°46′N, 0°26′E

On 'Orival', remains of a villa. 'Le Petit Moulin', a Gallo-Roman building of the Augustan period to about the end of 2nd c, reconstructed at the end of 3rd c; 17 rooms around a porticoed courtyard, mosaics, numerous objects.

Cochet, *S.I.,* 460: *Gallia* XXII (1964), 290; XXVI (1968), 369; XXVIII (1970), 276: *RSSHN* XL (1965), 79: *Forum* II (1972), 35.

COLOMBIÈRES Lug., *Baiocasses* (Calvados, F) 49°18′N, 0°58′W

Gallo-Roman foundations in the marsh of Monfréville; bricks, a figurine of Venus with two children in terracotta.

de Caumont, 1831, 235.

COMB END Brit., *Dobunni* (Gloucestershire, GB) 51°48′N, 2°01′W

Villa, corridor; mosaics, hypocausts, coins of late 4th c.

Arch IX (1789), 319–24; XVIII (1817), 112–13: Rainey, 1973, 57; RCHM, *Gloucestershire Cotswolds,* 1976, 35–6.

COMBE DOWN Brit., *Belgae* (Avon, GB) 51°21′N, 2°20′W

Villa, courtyard. Coins of 4th century. Dedication of '*principia*' by Naevius, an *adiutor procuratorum*, (early 3rd cent.) found re-used as coverstone of a sarcophagus (*RIB* 179).

VCH Somerset I (1906), 309–12.

COMBE ST NICHOLAS Brit., *Durotriges* (Somerset, GB) 50°53′N, 2°59′W

Villa, courtyard. Numerous mosaic pavements; coins of late 3rd c.

VCH Somerset I (1906), 333: Rainey, 1973, 58.

COMBLEY Brit., *Belgae* (Isle of Wight, GB) 50°41′N, 1°14′W

Villa; baths, mosaics, pottery and coins of 3rd c.

PIOWNHAS VI (1969), 271–82: Rainey, 1973, 58: *Britannia* VII (1976), 364–6.

COMMANA Lug., *Osismii* (Finistère, F) 48°25'N, 3°57'W

'Bois de la Roche' and 'Kerouat', villas, the former near road between Carhaix and l'Aberwrac'h, the latter beside that between Carhaix and Landerneau. Walls, traces of hypocausts. Coarse pottery and terra sigillata, at the former mid-1st c and 3rd and 4th c, at the latter 3rd c.

du Chatellier, 1907, 105: *BSAF,* 1971, 26.

COMPIÈGNE (Forêt de) Bel., *Suessiones* (Oise, F) 49°22'N, 2°53'E

Minor settlement, 'Mont-Berny'; 'Ville des Gaules', temple, baths; 'La Carrière (La Garenne du Roi)', baths. *Cf* Pierrefonds. 'St-Sauveur', Belgic pottery kiln, coin hoard Gordian III – Valerian II.

Grenier, *Manuel* IV.1 (1960), 337, 341: *Gallia* XXIX (1971), 226.

COMPTON Brit., *Atrebates* (Berkshire, GB) 51°30'N, 1°14'W

Two pottery kilns, 4th c.

TNFC VII (1934–37), 211–16.

COMPTON GROVE Brit., *Dobunni* (Gloucestershire, GB) 51°51'N, 1°56'W

Villa, corridor; hypocausts, pottery and coins chiefly of 3rd and 4th centuries.

JRS XXII (1932), 214–15.

CONDATE *cf* Condé-sur-Iton

CONDATE REDONVM *cf* Rennes

CONDATE SENONVM *cf* Montereau

CONDÉ-SUR-HUISNE Lug., *Carnutes*? (Orne, F) 48°23'N, 0°52'E

Villa.

MSAN IX (1835), 431.

CONDÉ-SUR-ITON Lug., *Aulerci Eburovices* (Eure, F) 48°50'N, 0°58'E

CONDATE VICVS.

It. A. 385.4.

Tab. Peut.

Minor settlement. 'Parigny', villa.

Desjardins, *Tab. Peut.,* 22: Poulain, G., *BSNÉP* XX (1912), 35–66: *BSPF* X (1913), 338–40.

CONDÉCOURT-GAILLON Lug., *Veliocasses* (Val-d'Oise, Yvelines, F) 49°02'N, 1°57'E

Minor settlement at 'La Coudraie' comprising numerous buildings of 1st, 2nd and 3rd c on both sides of the 'Chaussée Brunehaut', the road between Chartres and Beauvais.

BAVF I (1965), 29; II (1966), 18; III (1967), 27; IV (1968), 21; V (1969), 25: *Gallia* XXI (1963), 353; XXV (1967), 219; XXVIII (1970), 245: Roblin, 1971, 201.

CONFLANS-STE-HONORINE Lug., *Parisii* (Yvelines, F) 48°59'N, 2°06'E

CONFLVENTES (6th to 12th c).

Port at the confluence of the rivers Seine and Oise.

Comm S-O, 1884, 88–9: Toussaint, 1951, 68: Duval, 1961, 224–7, 284: Roblin, 1971, 10, 110 n 1, 178, 200.

CONGRESBURY Brit., *Belgae* (Avon, GB) 51°22'N, 2°48'W

Pottery kiln.

PSANHS CVIII (1963–64), 172–74.

CONSTANTIA *cf* Coutances

CONTALMAISON Bel., *Ambiani* (Somme, F) 50°01'N, 2°44'E

Gallo-Roman foundations.

Vasselle, 306.

CONTRES Lug., *Aulerci Cenomani* (Sarthe, F) 48°17'N, 0°29'E

Gallo-Roman foundations, mosaics, coins, 'Au Moulin'.

Bouton.

***CONTY** Bel., *Ambiani* (Somme, F) 49°44'N, 2°09'E

Minor settlement.

Agache-Bréart, 1975, 48: Agache, 1978, 163.

COOLING Brit., *Cantiaci* (Kent, GB) 51°27'N, 0°32'E

Salt-workings; evaporation vessels, pottery 1st and 2nd c.

KAR XXII (1970–71), 38–39: Nenquin, 1961, 80.

COOLING Brit., *Cantiaci* (Kent, GB) 51°26'N, 0°32'E

Twelve pottery kilns.

Arch Cant XLV (1933), xlii: *KAR* X (1967), 5–6.

COOPER'S HILL *cf* Great Witcombe

CORBEIL et ST GERMAIN-LÈS-CORBEIL Lug., *Parisii* (Essonne, F) 48°36'N, 2°29'E

Temple, habitations, cremation cemetery of 1st c and Christian cemetery.

Duval, 1961, 229, 234: Roblin, 1971, 7, 28, 183, 272, 325.

***CORBIE** Bel., *Ambiani* (Somme, F) 49°55'N, 2°30'E

Cemetery; amphorae, coins; at 'Les Templiers'.

Leduque, *Ambianie,* 70: Agache-Bréart, 1975, 48: Agache, 1978, 23, 25, 354, 364.

CORFE CASTLE *cf* Norden

CORFE MULLEN Brit., *Durotriges* (Dorset, GB) 50°47'N, 2°01'W

Pottery kiln; iron slag, coin of Caligula. Occupied from Iron Age to mid-4th c, the kiln itself in use *c* AD 40–50.

Ant J XV (1935), 42–55: RCHM *Dorset* II.3 (1970), 600.

CORIALLO, CORIALLVM *cf* Cherbourg

CORINIVM DOBVNNORVM *cf* Cirencester

CORIOSOLITES

Caes. *BG* II, 34; III, 7, 11; VII, 75.

Plin. *NH* IV, 107 (*Coriosvelites*).

Not. Gall. III, 7 (*civitas Coriosolitum*).

Not. Tiron. XCI, Z (*Coriosultas*).

CIL XIII.1, 616 (*c*(*ivis*) *Coriosolis*); notes, p 490; XIII.2, 8991, 8995, 9012 (milestones, *C*(*ivitate*) *Cor*(*iosolitum*)).

Armorican tribe included in Lugdunensis (later Lugdunensis III) with capital at FANVM MARTIS

(Corseul, *qv*).
Colbert de Beaulieu I, 126–30.

CORMAINVILLE Lug., *Carnutes* (Eure-et-Loir, F) 48°08′N, 1°36′E
Large villa near 'La Grande Pointe Mérou'.
Air photograph, Jalmain.

*CORMEILLES-EN-BEAUVAISIS Bel., *Bellovaci* (Oise, F) 49°39′N, 2°12′E
CORMILIACO, CVRMILIACA.
It. A. 380.2 (*Curmiliaca*).
Major settlement; frontier post ('la Neuville') 2 km to the north. Foundations, wells, tiles, coins, pottery; cemetery; Roman road between Amiens and Beauvais.
Agache, fig. 249: Agache-Bréart, 1975, 145: Agache, 1978, 423.

CORSEUL Lug., *Coriosolites* (Côtes du Nord, F) 48°29′N, 2°10′W
?FANVM MARTIS, later CIVITAS CORIOSOLITVM.
It. A. 387.1 (*Fano Martis*).
Tab. Peut.
Not. Dig. Occ. XXXVII, 19 (*praefectus militum Martensium, Aleto*).
CIL XIII, 3143–3147.
Wuilleumier, 339, 340.
Cf CORIOSOLITES.
Town, capital of the CIVITAS CORIOSOLITVM; important cross-roads. Founded in the Iron Age, flourished 2nd c, ended 5th c AD. Considerable foundations; temple. Pottery and coins from Gaulish to late Empire.
BSCN XXIX (1891), 193; XXXV (1897), 264; XLVII (1909), 54, 57–60: *Ann Bret* LXIV (1957), 116–9: Grenier, *Manuel* IV.2 (1960), 803–11: *Gallia* XXI (1963), 428; XXVII (1969), 248–51; XXIX (1971), 239–40; XXX (1972), 284–8: *Arche* XLVII (1972), 59–67.

COSEDIA *cf* Coutances

COUDRAY Lug., *Carnutes* (Eure-et-Loir, F) 48°40′N, 1°12′E
Cemetery.
Nouel, 27.

COULONCHE (La) Lug., *Aulerci Diablintes* (Orne, F) 48°39′N, 0°27′W
Coin hoard of the end of the 3rd c; moulds for coins of Diocletian, Maximinus, Constans.
BSAN VII (1875), 295; LIII (1955–56), 197–263.

COULONVILLERS Bel., *Ambiani* (Somme, F) 50°08′N, 2°00′E
Minor settlement 'le Festel', late Empire.
Agache, fig. 19, 508: *RN* XLIV (1962), 337: Agache-Bréart, 1975, 49.

*COURCELLES-SOUS-MOYENCOURT Bel., *Ambiani* (Somme, F) 49°49′N, 2°03′E
Villa. Small finds, coins.
Vasselle, 306: Agache-Bréart, 1975, 49.

COURCY-AUX-LOGES Lug., *Carnutes* (Loiret, F) 48°04′N, 2°13′E
Gallo-Roman foundations; coin hoard.

Nouel, 10, 21, 35.

COURGENARD Lug., *Aulerci Cenomani* (Sarthe, F) 48°09′N, 0°44′E
Temple, 'Mt. Chauvet'. Following the destruction of an ancient chapel dedicated to St Fiacre very substantial foundations were discovered, clearly representing the remains of the temple which was located here by local tradition and was dedicated to Jupiter. Gaulish and Roman coins.
Charles, R., *La station celtique du Crochemerlier*, Tours, 1875.

COURSEULLES Lug., *Viducasses* (Calvados, F) 49°20′N, 0°27′W
'Le Vieux Clos', 'les Fossettes', *vicus* or minor settlement, port. Stone-built coffins in cemetery at the 'camp de St Ursin'. Tiles, bricks, bronzes, amphora in the port, coarse pottery. Hoard of 4000 coins from Otacilia Severa to Quintillus, 3rd c.
de Caumont, 1831, 232; 1846, 370, 437: *MSAN* VI (1831–33), 70–74: *BSAN* XXIX (1913), 225–34, 422.

COUTANCES Lug., *Unelli* (Manche, F) 49°03′N, 1°27′W
COSEDIA, later CONSTANTIA.
It. A. 386.7; *Tab. Peut*. (*Cosedia*).
Not. Dig. Occ. XXXVII, 9 and 20 (*Constantia*).
Not. Gall. II, 8 (*civitas Constantia*).
'le Montcâtre', 'Chateau Pisquiny', town with port, on the road between Valognes and Rennes. Baths, villas, camp at Montcâtre, part of Gallo-Roman aqueduct, cement floors, burials. A cooking pot, a bucket, a bronze patera signed PVDESF: coins of Claudius, Nero, Vespasian, Domitian, Hadrian, Philip: Arretine ware, amphorae: objects of terracotta including head of divinity. Bronze vessel containing 237 coins from Vespasian to Gratian and with signature SVRIMA and an undoubtedly false description. Begun in 1st c, declined in 2nd and 3rd cc, ended 4th c.
de Gerville, 1854, 106: Guenault, 1862, *passim*: *BSAN* III (1865), 354–68; VII (1894–95), 194–99; LIII (1955–56), 197–263: *Ann 5 Norm*, 1886, 14–99: *BSAIC* XXIV (1900), 106–13; XXX (1908), 105: *BSNÉP* XIII (1905), 148–52: *Rev Manche* XVII (1963), 6–37: Johnson, 1979, 83.

COWBRIDGE Brit., *Silures* (South Glamorgan, GB) 51°28′N, 3°28′W
?BOVIVM.
It. A. 484.3 (*Bomium*).
JRS LVI (1966), 220 (brick stamp, BOV, now shown to be forgery).
PNRB 273.
Minor settlement.
Morgannwg XVII (1973), 59–60.

COWLEY Brit., ?*Catuvellauni* (Oxfordshire, GB) 51°43′N, 1°12′W
Pottery, many kilns, active 2nd to 4th c.
Oxon I (1936), 94–102; VI (1941), 9–21; XXXVIII (1973), 215–32.

COX GREEN Brit., *Atrebates* (Berkshire, GB) 51°30′N, 0°45′W

Villa, corridor; baths, hypocausts; late 2nd to late 4th c.

Berks A J LX (1962), 62–91.

CRACOUVILLE *cf* Vieil-Évreux

CRANBROOK Brit., *Cantiaci* (Kent, GB) 51°05′N, 0°32′E

Iron workings; quenching pond, furnaces, tiles some of which are stamped CL BR; window glass.

Arch Cant LXXII (1958), lx–lxii: *SxAC* CVII (1969), 109–11.

CRAYFORD Brit., *Cantiaci* (Greater London, GB) 51°27′N, 0°10′E

?NOVIOMAGVS.

It. A. 472.1 (*Noviomago*).
Tab. Peut. (. . . *madus*).
PNRB 428.

Major settlement on the road between London and Rochester. Foundations, cemetery.

Ant XXX (1956), 44: Rivet, 1964, 145: *Britannia* I (1970), 44, 77: Rodwell-Rowley, 1975, 62.

CREIL Bel., *Silvanectes* (Oise, F) 49°16′N, 2°29′E

Temple, by the roadside: villa Herbeval, fond de Malassise, quarries: villa de Houy, Forêt de la Haute-Pommeraie.

Gallia VII (1949), 112: Roblin, M., Le terroir de Creil, *Mém Soc hist arch Senlis,* LXIV–LXVI, 1968, 29–54: Durvin, P., *Archéologie Picardie Île-de-France,* 1, X–XII.

CRETTEVILLE Lug., *Unelli* (Manche, F) 49°21′N, 1°23′W

Foundations, possible villa; walls with brick courses, tegulae.

MSAN V (1829–30), LXII: de Gerville, 1854, 112: *BSAIC* XXIV (1900), 115.

CREUZY Lug., *Carnutes* (Loiret, F) 48°03′N, 1°52′E

Gallo-Roman foundations; statuary.

Nouel, 20, 29.

CRICKLADE Brit., *Dobunni* (Wiltshire, GB) 51°38′N, 1°51′W

RIB 100.

Minor settlement on the river Thames, possibly the river port of Cirencester; on the road between Cirencester and Silchester. Inhumations, coins and pottery of late Romano-British period.

WANHM LVI (1955), 162–66; LXVII (1972), 61–111.

CRIEL-SUR-MER Lug., *Caletes* (Seine-Maritime, F) 50°01′N, 1°19′E

(suff. -OIALVM corrupted to CRIOLIVM).

Gallo-Roman foundations, numerous burials both Gallo-Roman and Merovingian, brooches, weapons, scales, amber. At 'Mont-Jolibois, Côtes des Crocs', Gallo-Roman finds.

Leduque, *Ambianie,* 163, n 692–93.

CRIMBLEFORD *cf* Seavington St Mary

CRIQUEBEUF-SUR-SEINE Lug., *Aulerci Eburovices* (Eure, F) 49°18′N, 1°06′E

Minor settlement, temple, villa, at 'Le Câtelier'.

Vesly, 41: *BACTH,* 1898, 304: *Gallia* XXXVI (1978), 298–9.

CRIQUETOT-L'ESNEVAL Lug., *Caletes* (Seine-Maritime, F) 49°39′N, 0°17′E

Gallo-Roman foundations.

Cochet, *S.I.,* 349.

CRIQUIERS Lug., *Caletes* (Seine-Maritime, F) 49°41′N, 1°43′E

Gallo-Roman foundations at 'Authieux'.

Cochet, *S.I.,* 565, 592.

CROCIATONVM *cf* Carentan

CROCK HILL Brit., *Belgae* (Hampshire, GB) 50°56′N, 1°42′W

Pottery, many kilns making New Forest wares, late 3rd to early 4th c.

Detsicas, A. (ed), *Current Research in Romano-British Coarse Pottery,* London, 1973, 117–34.

CROISSY-SUR-CELLE Bel., *Ambiani* (Oise, F) 49°42′N, 2°11′E

Cemetery.

Leduque, *Ambianie,* 154.

CROIXDALLE Lug., *Caletes* (Seine-Maritime, F) 49°51′N, 1°22′E

Villa; large vaults beside the road to Londinières.

Cochet, *S.I.,* 543.

CROMHALL Brit., *Dobunni* (Avon, GB) 51°36′N, 2°27′W

Villa; mosaics, hypocausts.

PSAL 2 XXIII (1910), 20–23.

CROTOY (Le) Bel., *Ambiani* (Somme, F) 50°13′N, 1°37′E

Minor settlement on the beach and the coastal strip; port of the *Classis Sambrica.* Foundations, wells, tiles, small finds, statuette of Jupiter.

Agache, fig. 66–68, 297–98, 300: Leduque, *Ambianie,* 29 n. 54: *MSÉA* III (1836–1837), 208–09: Vasselle, 308: Agache, 1978, *passim.*

CROTTES-EN-PITHIVERAIS Lug., *Carnutes* (Loiret, F) 48°07′N, 2°04′E

Gallo-Roman foundations, wells, terra sigillata, statuary, querns, coin hoard.

Nouel, 22, 28, 29, 31, 35.

***CROUY** Bel., *Ambiani* (Somme, F) 49°58′N, 2°05′E

Villa.

Agache, fig. 110, 222, 352, 565: Leduque, *Ambianie,* 77 n. 352: Agache-Bréart, 1975, 49–51: Agache, 1978, 215, 252, 258, 327.

CROWHURST PARK FARM Brit., *Regni* (East Sussex, GB) 50°52′N, 0°30′E

Iron-working: furnaces, pottery, terra sigillata, tiles imprinted CL BR, tuyères.

Straker, 1931, 353: *SxAC* LXXIX (1938), 224–32.

CROZON Lug., *Osismii* (Finistère, F) 48°15′N, 4°29′W

At 'Lostmarc'h', cemetery with many hundreds of inhumations; head of a statuette and other small finds; hoard of minims of 4th c, coins of 3rd and 4th

c. At 'Kerzanvez', 'la Boixière' and 'Kervian', Gallo-Roman foundations, possibly villas. At the first, terra sigillata; at the last, a coin hoard of 3rd c. At 'l'Aber', on the Bay of Douarnenez, salt manufactory; terra sigillata of 2nd c.
BSAF, 1875, 126: *Bull Soc Brest*, 1903–04, 87–89: *Gallia* XXX (1972), 208.

CUIGNIÈRES Bel., *Bellovaci* (Oise, F) 49°28'N, 2°28'E
Gallo-Roman foundations along the Roman road, 2nd to 4th c. Merovingian cemetery.
Gallia XXIX (1971), 228.

CVNETIO *cf* Mildenhall

CURDRIDGE Brit., *Belgae* (Hampshire, GB) 50°54'N, 1°15'W
Tile kiln; coin of Trajan.
PHFC XXII (1961), 22.

*****CURLU** Bel., *Ambiani* (Somme, F) 49°58'N, 2°49'E
Temple, 'La Petite Chaudière'.
Agache, *BSAP,* 1972, 321: Agache-Bréart, 1975, 51.

*****CVRMILIACA** *cf* Cormeilles-en-Beauvaisis

CWM Y CADNO Brit., ?*Silures* (Dyfed, GB) 51°58'N, 3°44'W
Two practice camps.
JRS XLVIII (1958), 96: *Arch Camb* CXVII (1968), 119.

DAMBLAINVILLE Lug., *Viducasses* (Calvados, F) 48°55'N, 0°07'W
'Château Tarin', villa; walls, fragment of an altar of Vieux marble; coins from Nero to Constantine: 1st, 2nd, 4th c.
de Caumont, 1850, 440–44: *BSAN* XXIX (1913), 235; LII (1955), 110–11.

DAMBRON Lug., *Carnutes* (Eure-et-Loir, F) 48°07'N, 1°52'E
Gallo-Roman foundations, late.
Gallia XXX (1972), 316.

*****DAMÉRY** Bel., *Ambiani* (Somme, F) 49°44'N, 2°45'E
Gallo-Roman foundations (villa) 8 ha in extent; at the western exit, coins.
MSAP I (1838), 477: Leduque, *Ambianie,* 143 n. 294: Vasselle, 308: Agache-Bréart, 1975, 53: Agache, 1978, 44, 253, 438, 440, 444.

DAMPIERRE-EN-BRAY Lug., *Veliocasses* (Seine-Maritime, F) 49°32'N, 1°40'E
Villa. Traces of occupation near the church; 3rd-c coin hoard. Hamlet 'la Vieux-Ville'. At the hamlet of 'Campulay', a villa. The tradition of a vanished town.
Cochet, *S.I.,* 583.

DAMPMART Lug., *Meldi* (Seine-et-Marne, F) 48°53'N, 2°44'E
Cemetery, 'Champ Breton'; Gallo-Roman.
Gallia XXV (1967), 216.

DANCOURT Lug., *Caletes* (Seine-Maritime, F) 49°54'N, 1°33'E
Gallo-Roman foundations at the hamlet of St Rémy-en-Rivière; tiles. Poteau Saint-Rémy, 3rd-c coin hoard.
Cochet, *S.I.,* 547.

*****DAOURS** Bel., *Ambiani* (Somme, F) 49°54'N, 2°27'E
Villa; coins, small finds.
Leduque, *Ambianie,* 71 n. 291: Agache-Bréart, 1975, 53: Agache, 1978, 38, 339, 363.

DARENTH Brit., *Cantiaci* (Kent, GB) 51°24'N, 0°14'E
Villa, corridor; baths, frescoes, *fullonica*, hypocausts, mosaics.
Arch Cant XXII (1897), 49–84: *KAR* XIX (1970), 16: Rivet, 1969, 132–49.

DEMETAE
Ptol. II,3,12 (Δημῆται).
Gildas, 31.
Hist. Brit. II, 14, 47.
PNRB 333.
British tribe, occupying Dyfed (SW Wales) with capital at MORIDVNVM (Carmarthen, *qv*).

*****DEMUIN** Bel., *Ambiani* (Somme, F) 49°49'N, 2°33'E
Gallo-Roman foundations, ancient ditches; tiles.
Agache, fig. 477, 489: Leduque, *Ambianie,* 142: Vasselle, 308: Agache-Bréart, 1975, 53: Agache, 1978, 286, 287, 293, 314, 338.

DERCHIGNY-GRAINCOURT Lug., *Caletes* (Seine-Maritime, F) 49°56'N, 1°13'E
Gallo-Roman foundations; coins.
Cochet, *S.I.,* 254.

DERRY HILL Brit., *Belgae* (Wiltshire, GB) 51°26'N, 2°03'W
Temple, plan unknown; tiles, much pottery, many coins, late 3rd and early 4th c.
VCH Wiltshire I (1957), 54.

DESMONTS Lug., *Carnutes* (Loiret, F) 48°14'N, 2°30'E
Gallo-Roman foundations.
Nouel, 21.

DESVRES Bel., *Morini* (Pas-de-Calais, F) 50°40'N, 1°50'E
Gallo-Roman foundations.
Leduque, *Boulonnais,* 73.

DEUIL Lug., *Parisii* (Val d'Oise, F) 48°59'N, 2°20'E
Temple at 'Lac Marchais'.
Roblin, M., St-Eugène and the sacred lake of Deuil, *Paris et Île-de-France, Mém.,* 2, 7.

DEUX-JUMEAUX Lug., *Baiocasses* (Calvados, F) 49°21'N, 0°57'W
Small villa under the apse of the church. Terra sigillata PATERNVS, DOECCVS; fragments of Spanish amphorae; glass.
Gallia XX (1962), 420–21; XXII (1964), 284–86.

DEWLISH Brit., *Durotriges* (Dorset, GB) 50°46'N, 2°20'W

Villa, corridor; mosaics, painted plaster, hypocausts, coins and pottery of 4th c.
PDNHAS XCIV (1972), 81–86: Rainey, 1973, 60.

DIABLINTES cf *Aulerci*

DIEPPE Lug., *Caletes* (Seine-Maritime, F) 49°56′N, 1°04′E
Major settlement. Zone W, 'Côteau de la Caudecôte', buildings, cemetery (2nd–3rd c). Zone E, 'Bonne Nouvelle', buildings. Museum.
Frissard, P., *Notice historique sur Dieppe*, Paris, 1854: Cochet, *S.I.*, 235: *BCASM* VII, 195–307: Mangard, *Amis du Vieux Dieppe*, 1964, 53.

DIGULLEVILLE Lug., *Unelli* (Manche, F) 49°42′N, 1°51′W
Votive deposit under stones, including bronzes – a deer, a human mask, the trunk of a statuette wearing a sagum, a bull, a pedestal; an iron axe: a terra cotta patera and fragments of pots; five white clay figurines of Venus Anadyomene, a kidney in whitish chalcedony as big as a fist.
MSAN I, 2, (1824), 50–57: de Gerville, 1854, 112: *BSAIC* XXIV (1900), 38–9: Bouhier, 1962, *v* Digulleville.

DINGE Lug., *Redones* (Ille-et-Vilaine, F) 48°21′N, 1°43′W
Gallo-Roman foundations at 'Bourg'; traces of iron, coins. Bricks, querns, and a coin hoard of Antoninus Pius to Probus, 2nd to 3rd c, some 300 in all.
BSAIV V, 1861, 80: Orain, 1882, 103: *Bull Arch Ass Bret*, 1897, 231: Baucat, P., *Les voies romaines*, 57: Bizeul, *Des Coriosolites*, Dinan, 1858, 146–7.

DITCHEND cf Little Milton

DITCHLEY Brit., *Dobunni* (Oxfordshire, GB) 51°53′N, 1°25′W
Villa, corridor; enclosure. Two periods of occupation, from *c* AD 70 to 200 and from *c* 300 to 400. Tiles, window glass, pottery, painted plaster, coins.
Oxon I (1936), 24–69.

DIVODVRVM cf Jouars-Pontchartrain

DOBVNNI
Ptol. II,3,12 and 13 (Δοβούννοι/Δοβοῦνοι).
Cass. Dio LX,20,2 (τῶν Βοδούννων).
Rav. Cos. 106.31 (*Cironium Dobunorum*).
RIB 621, 2250.
CIL XVI, 49.
CIIC 428.
PNRB 339–40.
British tribe occupying the lower Severn valley and adjacent areas, with oppidum at Bagendon and capital at CORINIVM DOBVNNORVM (Cirencester, *qv*).
Clifford, E. M. (ed.), *Bagendon: a Belgic Oppidum*, Cambridge, 1961: Rivet, 1964, 151–3: Frere, 1978, 66–7: Webster, 1980, 60–61 etc.: McWhirr, A., *Roman Gloucestershire*, Gloucester, 1981, 2–3.

DOLVCENSIS VICVS cf ?Halinghen, ?Isques

DOMART-SUR-LA-LUCE Bel., *Ambiani* (Somme, F) 49°49′N, 2°29′E
Gallo-Roman foundations; pottery.
Leduque, *Ambianie*, 142: Agache-Bréart, 1975, 54.

DOMFRONT-EN-CHAMPAGNE Lug., *Aulerci Cenomani* (Sarthe, F) 48°06′N, 0°01′E
?Fort. On the rise known as the 'Grand-Gagné' an enclosure known as 'Camp de César' formed by a rampart with external ditch, and with a single entrance. Jaillot's map indicates the ruins. In 1815 there were discovered coins and a Constantinian ring with the Christian symbol.
Annuaire du dépt. de la Sarthe, 1815, 30; 1834, 59–79: Pesche, *Dictionnaire* II, 77, 222–4: Voisin, 1862, 68: Ledru, 1911, 76.

DOMPIERRE-DU-CHEMIN Lug., *Redones* (Ille-et-Vilaine, F) 48°16′N, 1°09′W
Gallo-Roman foundations.
Pauthiel, 1927, 23.

***DOMPIERRE-EN-SANTERRE** Bel., *Ambiani* (Somme, F) 49°55′N, 2°48′E
Temple, at 'Sole de Cany'.
Agache, fig. 217; 1972, 321: Agache-Bréart, 1975, 54: Agache, 1978, 312, 391, 393.

DOMPIERRE-SUR-AUTHIE Bel., *Ambiani* (Somme, F) 50°18′N, 1°55′E
Temple at 'Bois du Préel'.
Agache, fig. 424; 1972, 322: Agache-Bréart, 1975, 54: Agache, 1978, 245, 391.

DOMQUEUR Bel., *Ambiani* (Somme, F) 50°07′N, 2°03′E
DVRO(I)CO REGVM.
Tab. Peut.
Major settlement on the road between Boulogne and Amiens. Gallo-Roman foundations, bridge, coin hoard, amphorae.
Bastien, P., and Vasselle, F., *Le trésor monétaire de D.*, Wetteren, 1965: Leduque, *Ambianie*, 128: Ponchon, A., Le pont de D., *BSAP* XXIX (1923), 298–307: *BACTH* XL (1922), cxiv–cxv: Vasselle, 308: Agache-Bréart, 1975, 54.

DOMVAST Bel., *Ambiani* (Somme, F) 50°12′N, 1°55′E
Gallo-Roman foundations, coins.
Agache, fig. 180: Leduque, *Ambianie*, 175, 893: Vasselle, 330: Agache-Bréart, 1975, 54.

DORCHESTER Brit., *Catuvellauni* (Oxfordshire, GB) 51°39′N, 1°10′W
RIB 235.
Major settlement at confluence of rivers Thames and Thame and on the road between Silchester and Alchester. The name TAMESE (*Rav. Cos.* 106.38) has been identified with Dorchester, but there is no support for this and perhaps a stronger case for identifying TAMESE with the river Thames, *qv*. Walls enclose *c* 5.5 ha. Cemetery. Occupation extends from Iron Age, Romano-British and early Saxon settlements to the present day. Pottery kiln.
Oxon I (1936), 81–94: *Arch J* CXIX (1962), 114–49: *JRS* LIV (1964), 166: *Britannia* IV (1973), 297: Rodwell-Rowley, 1975, 115–118.

DORCHESTER Brit., *Durotriges* (Dorset, GB)

50°43'N, 2°26'W
DVRNOVARIA.
It. A. 483.6, 486.15 (*Durnonovaria*).
?*Rav. Cos.* 105.22 (*Duriarno*).
RIB 188–90.
PNRB 345.

Town, capital of the CIVITAS DVROTRIGVM (but *cf* Ilchester). On the river Frome, with roads to Ilchester, Exeter and Badbury (*Vindocladia*). Walls, enclosing *c* 30 ha; aqueduct; large cemetery at Poundbury; amphitheatre in converted henge monument, Maumbury Rings; many small finds. Occupied from *c* AD 70 to the early 5th c. Museum.

PDNHAS, passim: RCHM *Dorset* II.3 (1970), 531–92: Rainey, 1973, 61–7: Wacher, 1975, 315–326.

DOUARNENEZ-PLOARE-TRÉBOUL Lug., *Osismii* (Finistère, F) 48°06'N, 4°21'W
Wuilleumier p 136, no 338.

Town on the Bay of Douarnenez, junction of numerous roads towards Quimper, Audierne, the neighbourhood of Crozon, etc. Many foundations; burned burials, a lead sarcophagus. Coins from Augustus to Constantine. Much local and imported pottery, terra sigillata from Lezoux; Argonne 4th c. Cologne glass. Many salt manufactories in and around the town.

du Chatellier, 1907, 255–58: *Gallia* XXX (1972), 209–13: *Ann Bret* LXXX (1973), 215–36.

DOULLENS Bel., *Ambiani* (Somme, F) 50°09'N, 2°21'E
Aqueduct; wells, coins, small finds, ?vicus (Fossier, R.).

Leduque, *Ambianie*, 131: Agache-Bréart, 1975, 54: Agache, 1978, 21, 38, 40, 41, 208.

DOURGES *cf* Noyelles-Godault

DOVER Brit., *Cantiaci* (Kent, GB) 51°07'N, 1°19'E
PORTVS DVBRIS.
?Ptol. II,3,3 (Καινὸς λιμήν).
It. A. 473.2, 473.5 (*ad portum Dubris*).
Tab. Peut. (*Dubris*).
Not. Dig. Occ. XXVIII, 4 and 14 (*Praepositus militum Tungrecanorum, Dubris*).
Rav. Cos. 106.35 (*Dubris*).
PNRB 341, 428.

Fort and port at the mouth of the river Dour (DVBRIS FLVMEN: *Rav. Cos.* 108.38). 2nd c fort of the *Classis Britannica*, probably its British headquarters. Walls, barracks, granary, many other buildings; numerous tiles stamped CL BR. Extra-mural settlement, including the 'Painted House' (built *c* 200) with frescoed walls in magnificent state of preservation. 3rd c fort of the *Litus Saxonicum*. Cemetery, two lighthouses, museum.

Arch XCIII (1949), 32: Cunliffe, 1968, 258: *Arch J* CXXVI (1969), 78–100: Collingwood-Richmond, 1969, 66: *SxAC* CVII (1969), 102–25: *KAR* XXIII (1971), 76–86: Philp, B. J., *Roman Dover*, D., 1973: Johnston, 1977, 20–21: Johnson, 1979, 51–3; Philp, B. J. *The Excavation of the Roman Forts of the Classis Britannica at Dover 1970–71*, Dover, 1981.

DOWNTON Brit., *Belgae* (Wiltshire, GB) 50°59'N, 1°45'W
Villa, corridor; baths, mosaics. Founded late 3rd c, flourished early 4th and declined mid-4th c.

WANHM LVIII (1961–3), 303–41: Rainey, 1973, 67.

DRAVEIL Lug., *Parisii* (Essonne, F) 48°41'N, 2°25'E
Gallo-Roman foundations; amphorae, weapons, roof tiles, pruning knife.
Duval, 1961, 231: Roblin, 1971, 8, 174, 328.

DREUX Lug., *Carnutes* (Eure-et-Loir, F) 48°44'N, 1°22'E
DVROCASSES.
It. A. 384.5, 385.1, 385.5 (*Durocasis*).
Tab. Peut.
Major settlement.
Duval, 1961, 118, 244–5.

DROCOURT Bel., *Atrebates* (Pas-de-Calais, F) 50°23'N, 2°55'E
Villa.
Dérolez, 514.

DROPSHORT Brit., *Catuvellauni* (Buckinghamshire, GB) 51°59'N, 0°42'W
MAGIOVINIVM.
It. A. 471.1 (*Magiovinto*), 476.10, 479.6 (*Magiovinio*).
PNRB 406.

Minor settlement on Watling Street following early fort between Dunstable and Towcester. Defensive rampart, pottery and coins of 3rd and 4th c.

VCH Buckinghamshire II (1908), 4–5: *Britannia* II (1971), 268; *Records of Bucks.* XX, iii (1977), 384–99.

DRUCAT Bel., *Ambiani* (Somme, F) 50°09'N, 1°52'E
Gallo-Roman foundations; cemetery.
Agache, fig. 41, 48, 153, 171, 279, 311, 331, 333: Leduque, *Ambianie*, n. 291: Agache-Bréart, 1975, 55: Agache, 1978, *passim*.

DRYHILL Brit., *Dobunni* (Gloucestershire, GB) 51°51'N, 2°06'W
Villa, corridor; baths, hypocausts, coins of 3rd and 4th c.

TBGAS IV (1879–80), 208; RCHM, *Gloucestershire Cotswolds*, 1976, 5–6.

DVBRIS PORTVS *cf* Dover

DUISANS Bel., *Atrebates* (Pas-de-Calais, F) 50°18'N, 2°41'E
Temple, 'La Devalonne, Derrière las Haies'.
Agache, 1972, 322: Leduque, *Atrébatie*, 62: Agache-Bréart, 1975, 137: Agache, 1978, 400.

DVMNONII
Ptol. II,3,13 (Δουμνόνιοι/Δαμνόνιοι).
?Solinus XXII, 7.
It. A. 483.8, 486.8, 486.17 (*Isca Dumnoniorum*).
Tab. Peut. (*Iscadumnoniorum*).
Gildas XXVIII.
Rav. Cos. 106.2 (*Scadumnamorum*), 106.6 (*Scadoniorum*).

RIB 1843 (*civitas Dum(no)ni(orum)*), 1844 (*civitas Dumnoni(orum)*).
PNRB 362–3.
British tribe occupying Devon and Cornwall, with capital at ISCA DVMNONIORVM (Exeter, *qv*).
Rivet, 1964, 153–5: Fox, A., *South-West England*, Newton Abbot, 1973.

DVMNONIVM SIVE OCRINVM PROMONTORIVM *cf* Lizard (The)

DUNDRY DOWN Brit., *Belgae* (Avon, GB) 51°24′N, 2°38′W
Stone quarry.
Hebditch, M and Grinsell, L. V., *Roman Sites in the Mendips*, Bristol, 1968, 26.

DUNEAU Lug., *Aulerci Cenomani* (Sarthe, F) 48°04′N, 0°41′E
Villa.
Bouton.

DUNSTABLE Brit., *Catuvellauni* (Bedfordshire, GB) 51°53′N, 0°31′W
DVROCOBRIVIS.
It. A. 471.2, 476.9, 479.7 (*Durocobrivis, Durocobrius*).
PNRB 349.
Minor settlement on Watling Street between St Albans and Towcester. Timber buildings, pottery, coins.
VCH Bedfordshire II (1908), 6: *Britannia* II (1971), 267: Matthews, C. L., *Ancient Dunstable*, Dunstable, 1963, 55–66.

DVNVM *cf* Châteaudun, ?Hod Hill
DVRNOVARIA *cf* Dorchester (Dorset)
DVROBRIVAE *cf* Rochester
DVROCASSES *cf* Dreux
DVROCOBRIVIS *cf* Dunstable
DVROCORNOVIVM *cf* Wanborough
DVROICOREGVM *cf* Domqueur
DVROLEVVM *cf* ?Ospringe
DVROLITVM *cf* ?Chigwell, ?Romford
DVROTRIGES
Ptol. II,3,13 (Δουρότριγες).
RIB 1672 (*c(ivitas) Dur(o)tr(i)g(um) [L]endin(i)e(n)sis*), 1673 (*ci(vitas) Durotrag(um) Lendinie(n)si[s]*).
PNRB 352–3.
British tribe, occupying Dorset and Somerset, with numerous oppida including Ham Hill, Hod Hill and Maiden Castle. One of the two *validissimae gentes* conquered by Vespasian, AD 43 ff (Suetonius, *Vesp.* 4, 1; the identity of the other *gens* is disputed). Established as a *civitas* with capital at DVRNOVARIA (Dorchester, Dorset, *qv*); under the late Empire the northern part of this may have been separated and given a capital at LINDINIS (Ilchester, *qv*).
Rivet, 1964, 155–6: Wacher, 1975, 407–8: Branigan-Fowler, 1976, 15–27 *passim*: Frere, 1978, 67–8: Webster, 1980, 61, 108–10: Salway, 1981, 92–3.

DVROVERNVM CANTIACORVM *cf* Canterbury

DYMCHURCH Brit., *Cantiaci* (Kent, GB) 51°01′N, 0°59′E
Salt workings; briquetage, pottery, terra sigillata, 1st–2nd c.
Nenquin, 1961, 82: *SxNQ* XVII (1968), 23–24.

DYMOCK Brit., *Dobunni* (Gloucestershire, GB) 51°59′N, 2°26′W
Minor settlement based on iron working, occupied 2nd and 3rd c.
Britannia I (1970), 293: McWhirr, 1981, 67–70.

EASNEYE WOOD Brit., *Trinovantes* (Hertfordshire, GB) 51°48′N, 0°00′E
Tumulus; 18 m diameter at base, 3 m high.
PSAL XVII (1899–1901), 8.

EAST ANTON Brit., *Belgae* (Hampshire, GB) 51°13′N, 1°28′W
?LEVCOMAGVS.
Rav. Cos. 106.21 (*Leucomagno*).
PNRB 389.
Minor settlement at crossing of roads between Silchester and Dorchester (Dorset) and Cirencester and Winchester. Foundations of buildings, rubbish pits, pottery.
Britannia II (1971), 282; III (1972), 348.

EASTBOURNE Brit., *Regni* (East Sussex, GB) 50°46′N, 0°16′E
Salt workings; briquetage, pottery.
SxNQ XVII (1968), 23–24.

EASTBOURNE Brit., *Regni* (East Sussex, GB) 50°46′N, 0°17′E
Villa, corridor; bath, mosaics. Probably dating from 1st c.
VCH Sussex III (1935), 24: Cunliffe, 1973, 78–9.

EAST COKER Brit., *Durotriges* (Somerset, GB) 50°55′N, 2°39′W
Villa; walls, mosaics, hypocausts, coins mostly 4th c.
VCH Somerset I (1906), 329–31: Rainey, 1973, 68.

EAST GRIMSTEAD Brit., *Belgae* (Wiltshire, GB) 51°03′N, 1°40′W
Villa, corridor; baths, coins mostly of late 3rd c.
VCH Wiltshire I (1957), 75.

EAST MALLING Brit., *Cantiaci* (Kent, GB) 51°17′N, 0°27′E
Villa, type unknown; mosaics, murals.
Arch Cant LXXX (1965), 257.

EASTON GREY Brit., *Dobunni* (Wiltshire, GB) 51°35′N, 2°09′W
Minor settlement, 'White Walls', at crossing of the river Avon by the road between Cirencester and Bath, the Fosse Way. The possible abutments of a bridge are recorded. Coins from Julius Caesar to Constantine I and pottery from 1st to 4th c indicate long occupation. Foundations, mosaics. Relief of a goddess with three gods. Finds in Malmesbury museum.

VCH Wiltshire I (1957), 68: Ross, 1967, 211.

EAST TILBURY Brit., *Trinovantes* (Essex, GB) 51°28′N, 0°26′E

Salt workings; briquetage, evaporation vessels.
Nenquin, 1961, 90: *VCH Essex* III (1963), 190.

EBVROVICES cf *Aulerci*

ECCLES Brit., *Cantiaci* (Kent, GB) 51°19′N, 0°28′E

Large villa, corridor, 1st–4th c., on site occupied in Iron Age and succeeded by Anglo-Saxon cemetery; baths, mosaics, tile kiln.
Arch Cant LXXXVIII (1963) – XCI (1976), passim: Jarrett-Dobson, 1965, 105–8.

*ÉCHELLE (L')-SAINT-AURIN Bel., *Ambiani* (Somme, F) 49°42′N, 2°43′E

Cemetery of the Gallo-Roman period at the hamlet of Blamont.
Leduque, *Ambianie*, 143: Agache-Bréart, 1975, 57.

ÉCHILLEUSES Lug., *Carnutes* (Loiret, F) 48°10′N, 2°26′E

Gallo-Roman foundations.
Nouel, 21.

ÉCOUST-ST-MEIN Bel., *Atrebates* (Pas-de-Calais, F) 50°11′N, 2°55′E

Large villa, 'Le Buisson-St.-Mein, L'Homme Mort'.
Agache, *BCDMHPC* IX.1 (1971), 49: Agache-Bréart, 1975, 137–38: Agache, 1978, 324, 441.

ÉCULLEVILLE Lug., *Unelli* (Manche, F) 49°41′N, 1°49′W

'L'Hague-Dike', Gallo-Roman foundations near the mill and on the shore; tiles, bricks, 14 furnaces filled with cinders.
Et G et H Manche, 1854, 113: *BSAIC* XXIV (1900), 39.

EDINGTON Brit., ?*Durotriges* (Somerset, GB) 51°11′N, 2°54′W

Salt working; pottery of 3rd and 4th c.
VCH Somerset I (1906), 353.

EGLISTON Brit., *Durotriges* (Dorset, GB) 50°38′N, 2°09′W

Shale working.
RCHM *Dorset* II.3 (1970), 613.

ÉGREVILLE Lug., *Senones* (Seine-et-Marne, F) 48°10′N, 2°52′E

Gallo-Roman foundations.
Nouel, 21.

ÉGRY Lug., *Senones* (Loiret, F) 48°06′N, 2°26′E

Cemetery.
Nouel, 27.

ÉLETOT Lug., *Caletes* (Seine-Maritime, F) 49°47′N, 0°27′E

Gallo-Roman foundations, 'côte des Vaguans'.
BCASM XVIII, 3, 88.

ELING Brit., *Atrebates* (Berkshire, GB) 51°28′N, 1°14′W

Villa; mosaics, hypocausts, coin of Constantine I.
VCH Berkshire I (1906), 210.

ELSTREE Brit., *Catuvellauni* (Hertfordshire, GB) 51°39′N, 0°18′W

Tile kiln, late 1st and early 2nd c.
TLMAS XVI (1951), 229–33.

ELY Brit., *Silures* (South Glamorgan, GB) 51°29′N, 3°13′W

Villa; baths; occupied early 2nd to late 3rd c.
JRS XI (1921), 67–85.

ENCOMBE Brit., *Durotriges* (Dorset, GB) 50°37′N, 2°05′W

Shale working; pottery; late 2nd c.
RCHM *Dorset* II.3 (1970), 599.

ENGLISH CHANNEL – LA MANCHE
OCEANVS BRITANNICVS, MARE BRITANNICVM.

Str. II,5,28 (C.128) (τῷ Πρεττανικῷ πορθμῷ).
Mela I, 15; II, 85 (*Britannicum oceanum*); III, 48 (*in Britannico mari*).
Plin. *NH* IV, 109; VII, 206 (*Britannicus oceanus*).
Ptol. II,3,3; 8,2; 9,1; VIII,3,2; 5,2 (Πρεττανικὸς Ὠκεανός).
Pseudo-Agathemerus XIV (*GGM* II, 506) (ὠκεανὸς Βρεττανικός).
Tert. *de Cultu Feminarum* 1,6,2 (*Britannicum mare*).
Cass. Dio LIII,12,6 (ὠκεανοῦ τοῦ Βρεττανικοῦ).
Eutropius VI,17,2 (*ad oceanum Brittanicum*).
Amm. Marc. XXIII,6,88 (*in Brittanici secessibus maris*).
Hier. *in Gen.* X, 4 (*oceanum Britannicum*).
Dimensuratio Provinciarum XXX (*GLM*) (*Oceano Britannico*).
Marcian. II,24; II,44 (Πρεττανικὸς ὠκεανός).
Oros. I,2,63 (*oceanum Brittanicum*).
Iul. Hon. XV (*mare Britannicum*).
Vib. Seq. *Flumina*, 98 (*Occeanum Brittanicum*; ?a late interpolation).
Sozomenus I,7,2 (Βρεττανοὺς θαλάσσης).
Cassiod. I,9,2 (*Brittanicum mare*).
Ven. Fort. *Vita S Albini* V, 11 (*oceano Britannico*).
Rav. Cos. pp 322, 325 (*oceanum Britanici*); 344 (*oceanum Britannicum*).
CIL III, 247. (*ex oceano Britannico*).
PNRB 44.

(*NB* – Some of these seem to refer generally to 'the sea around Britain' rather than specifically to the Channel. Thus Tertullian and Ammianus write of pearls, probably from the North Sea coast of Kent or Essex).

ENTRAMMES Lug., *Aulerci Diablintes* or *Arvii* (Mayenne, F) 48°00′N, 0°43′W

Temples at 370100/330750 and 370350/330750, one at each.

ÉPIAIS-RHUS Lug., *Veliocasses* (Val d'Oise, F) 49°07′N, 2°04′E

Temple. On the plateau a sanctuary site over 40 ha, comprising a principal sanctuary probably with a portico and several temples in the Celtic tradition revealed by aerial photography. Baths (?) and a habitation zone with a cemetery.

Gallia XV (1957), 164; XIX (1961), 295; XXX (1972), 309: *BAVF* V (1969), 58–86, 87–102; VI (1970), 69.

ÉPINAY-SUR-SEINE Lug., *Parisii* (Seine-Saint-Denis, F) 48°57′N, 2°19′E
Gallo-Roman foundations; wells, Roman road, bridge, pottery.
RA XVI.2 (1860), 615–27: Duval, 1961, 231, 234: Roblin, 1971, 29, 166.

ÉPISY Lug., *Senones* (Seine-et-Marne, F) 48°20′N, 2°47′E
Gallo-Roman foundations.
Nouel, 21.

EPPING Brit., *Trinovantes* (Essex, GB) 51°41′N, 0°07′E
Tile kiln.
TEAS IV (1893), 222–23.

EPSOM Brit., ?*Regni* or *Atrebates* (Surrey, GB) 51°20′N, 0°18′W
Tile kiln, in use *c* AD 70–150.
SyAC XLV (1937), 90–92.

ÉQUIQUEVILLE Lug., *Caletes* (Seine-Maritime, F) 49°49′N, 1°26′E
Gallo-Roman foundations, tiles, the siding of 'Châtelets'.
Cochet, *S.I.,* 309.

ERGNIES Bel., *Ambiani* (Somme, F) 50°05′N, 2°02′E
Gallo-Roman foundations, tiles, small finds.
Vasselle, 312: Agache-Bréart, 1975, 57–58: Agache, 1978, 438, 453.

ERMONT Lug., *Parisii* (Val d'Oise, F) 48°59′N, 2°16′E
Gallo-Roman foundations; coin hoard, cemetery, early and late Empire.
Duval, 1961, 232, 247, 278: *Gallia* XXV (1967), 219: Roblin, 1971, 31, 122, 156, 241.

ERONDELLE *cf* Liercourt

ERQUY Lug., *Coriosolites* (Côtes du Nord, F) 48°38′N, 2°28′W
REGINEA?
Tab. Peut.
'le Pussover', on the road between Corseul and Erquy, numerous Gallo-Roman foundations, aqueduct, buildings, ?temple. Fibula; series of coins to about the middle of the 4th c.
MSCN I (1883–4), 118: Blanchet, 1908, 104: Grenier, *Manuel* IV.2 (1960), 811–3.

ESLETTES Lug., *Caletes* (Seine-Maritime, F) 49°33′N, 1°03′E
Cemetery.
Cochet, *S.I.,* 193.

ESQUIBIEN Lug., *Osismii* (Finistère, F) 48°01′N, 4°32′W
'Sugensou', villa beside the river Goyen and the road between Douarnenez and Audierne. Foundations, paving; Lezoux pottery from end of 2nd c.
BSAF, 1970, 48.

ESSARTS (Les) *cf* Rouvray (fôret de)

ESTAIRES Bel., *Menapii* (Nord, F) 50°38′N, 2°43′E
?MINARIACVM.
It. A. 377.4, 377.7.
CIL XIII, 9039.
Major settlement. Milestone on the road between Arras and Cassel, Trebonianus Gallus and his son. Five bronze statues in 'la Lys'.
Leduque, *Atrébatie,* 77: *Congrès Féd. hist. Belgique,* XXXIX, 1966.

ESTREBOEUF Bel., *Ambiani* (Somme, F) 50°09′N, 1°37′E
Gallo-Roman foundations, small finds, Roman road.
Agache, fig. 89, 233: Leduque, *Ambianie,* 165, n.737: Agache-Bréart, 1975, 59: Agache, 1978, 255.

ESTRÉE-LÈS-CRÉCY Bel., *Ambiani* (Somme, F) 50°15′N, 1°55′E
Gallo-Roman foundations, 'Bois d'Aouste' (Augusta?). Ditches.
Agache, fig. 43, 44, 47, 136, 356: Leduque, *Ambianie,* 128: Agache-Bréart, 1975, 59.

*ESTRÉES-SUR-NOYE Bel., *Ambiani* (Somme, F) 49°47′N, 2°20′E
Temple, 'Les Coutures'. Late La Tène pottery.
Agache, fig. 367, 396, 401, 417, 418, 599; 1972, 322: Leduque, *Ambianie,* 149: Agache, 1978, *passim.*

ESVVII *cf Sagii*

*ÉTALON Bel., *Ambiani* (Somme, F) 49°45′N, 2°51′E
Two villas, W of the village and N of the Bois de Liancourt. 'Bois d'Étalon, Les Terres Noires, le Paraclet'; villa with mosaics. 'Bois des Boeufs'; 3rd-c villa with hypocausts, cellar, small finds.
Duhamel-Drujean, *Description du canton de Nesle,* Peronne/St-Quentin, 1884, 50: *Gallia Supp* XIII (1959), 98; XXIX (1971), 231: Vasselle, 314: Agache, fig. 498: Agache-Bréart, 1975, 60.

ÉTAMPES Lug., *Carnutes* (Essonne, F) 48°26′N, 2°09′E
STAMPAE.
Greg. Tur. *HF* IX, 20; X, 19.
Chron. de Fredeg., a. 603–604.
Gallo-Roman foundations; ?large villa, 'Ville Sauvage', 'La Treille'; Merovingian coins. Municipal museum.
Gallia XXIII (1965), 327: Jalmain, 73: Nouel, 3, 5, 7, 21.

ÉTAPLES-SUR-LA-QUANTIN Bel., *Morini* (Pas-de-Calais, F) 50°31′N, 1°39′E
?PORTVS AEPATIACI (*qv*).
QVENTAVIC.
Beda *Hist Ecc* IV,1 (*Quaentavic*).
Small settlement and naval base of the *Classis Sambrica* at the mouth of the river Canche ('La Pièce aux Liards', 'Motte des Crouquelets'). Dwellings 2nd and 3rd c, cemetery 2nd c, well, kiln, small finds, terra sigillata, statuette, numerous fibulae (?workshop), hoard of 5000 coins, chiefly Postumus. Museum.
BSNAF, 1897, 338–50: Leduque, *Boulonnais,* 92

n 29; *Morinie,* 70: *Gallia* XXV (1967), 210; XXVII (1969), 228; XXIX (1971), 229: *Septentrion* I (1969), 16: Couppé, J., *Catalogue du Musée Quentovic,* Étaples, 1972: *BCDMHPC* IX (1973), 119, 126.

*ÉTINEHEM Bel., *Ambiani* (Somme, F) 49°56'N, 2°41'E

Gallo-Roman foundations (?temple) at 'Champ Bailly'.

Agache, 1972, 322: Leduque, *Ambianie,* 69 n 242: Agache-Bréart, 1975, 61: Agache, 1978, 40.

*ÉTOILE (L') Bel., *Ambiani* (Somme, F) 50°02'N, 2°02'E

Minor settlement. 'Camp de César, le Catelet', oppidum of the *Ambiani.* Temple ('les Coutures'), well, small finds, marble statuette; Gallic enclosure reoccupied by a villa. At 'Moreaucourt', sarcophagus, tombs. At 'Le Longuet', Gallo-Roman foundations.

BSPF XVII (1920), list lxxix, 56: Grenier, *Manuel* I.1 (1931), 193 n 4: *RN,* 1962, 323: Agache, fig. 90–91, 117, 145, 154–5, 374, 735: *BSPN* VI (1963–4): *BSAP,* 1972, 322: Leduque, *Ambianie,* 78: Agache-Bréart, 1975, 61.

ÉTRETAT Lug., *Caletes* (Seine-Maritime, F) 49°43'N, 0°12'E

Villa, in the old presbytery; urns, chapel of St-Nicolas. Aqueduct, 'Fond du Petit Val'. Cremation cemetery at 'bois des Maules'.

Cochet, *S.I.,* 358–59; *Étretat,* Dieppe, 1869.

ÉTRUN Bel., *Atrebates* (Pas-de-Calais, F) 50°19'N, 2°42'E

Gallo-Roman foundations. 'Camp de César', oppidum of the *Atrebates.*

BSPF XIV (1917), list LXIII, 476: Jelski, 142.

EU Bel., *Ambiani*/Lug., *Caletes* (Seine-Maritime, F) 50°03'N, 1°25'E

Minor settlement. Roads to Lillebonne, Beauvais, Amiens and Boulogne. Cemetery and small finds at the foot of 'Mont Blanc'. In the forest of Eu a temple ('Le Bois l'Abbé', 'Le Vert Ponthieu') dominating a height at the border between the *Ambiani* and the *Caletes;* cella, gallery, 2nd to 4th c, numerous small finds; obliterated enclosure, theatre ('La Cirque'), possible glass-workers' workshops.

Cochet, *S.I.,* 320: Grenier, *Manuel* IV.2 (1960), 747–8: Agache, fig. 17, *cf* 10, 96: *RSSHN* XL (1965), 73–6: *Gallia* XXIV (1966), 268; XXVI (1968), 370; XXVIII (1970), 276; XXX (1972), 345: Leduque, *Ambianie,* 169: *Ann Ass Norm* XXXIX, 437.

EURE, river Lug., *Carnutes, Aulerci Eburovices* (Eure-et-Loir, Eure, F)

AVTVRA FLVMEN.

Hence AVTRICVM (Chartres, *qv*).

ÉVREUX Lug., *Aulerci Eburovices* (Eure, F) 49°01'N, 1°09'E

MEDIOLANVM AVLERCORVM, later EBVROVICES, EBROICIS.

Ptol. II,8,9 (Μεδιολάνιον).

It. A. 384.4 (*Mediolano Aulercorum*), 384.13 (*Mediolanum*).

Tab. Peut. (*Mediolano Aulercorum*).

Amm. Marc. XV,11,12.

CIL XIII.1, p 510.

Cf AVLERCI.

Town, replacing Vieil-Évreux (*qv*) as capital of the CIVITAS EBVROVICVM mid 1st c. Port on the river Iton, road junction, emporium, agricultural market, industrial city (fullers, potters, metal-workers, 1st c). Forum, theatre (Claudian), fortifications (3rd c), cemetery. Museum.

Blanchet, *Enceintes,* 37–8: Espérandieu IV (1911), nos 3060–70: Grenier, *Manuel* I (1931), 424; III.2 (1958), 950–52: Duval, 1961, 236, 245: *Gallia* V (1947), 450; VII (1949), 122; IX (1951), 84; XX (1962), 423; XXIV (1966), 260; XXX (1972), 342.

EWELL Brit., ?*Regni* or *Atrebates* (Surrey, GB) 51°21'N, 0°15'W

Large settlement, on Stane Street between London and Chichester. Occupied c. 200 BC to AD 150 and then again in late 4th c. Buildings, ritual shafts, pottery, coins.

VCH Surrey IV (1912), 361–63: *SyAC* L (1947), 9–46; LXIX (1973), 1–26: Coles, J. and Simpson, D. A. A. (edd.) *Studies in Ancient Europe,* Leicester, 1968, 284: Rodwell-Rowley, 1975, 61.

EWENNY, river Brit., *Silures* (Mid Glamorgan, GB)

AVENTIVS FLVMEN.

Rav. Cos. 108.28 (*Aventio*).

PNRB 260–261.

EWHURST Brit., *Regni* (Surrey, GB) 51°10'N, 0°27'W

Villa; Flavian timber house rebuilt *c* AD 200, burned *c* 350. Baths, mosaics.

SyAC LXV (1968), 1–70.

EXE, river Brit., *Dumnonii* (Devon, GB)

ISCA FLVMEN.

Ptol. II,3,3 (Ἴσκα ποταμοῦ ἐκβολαί).

PNRB 376–8.

EXETER Brit., *Dumnonii* (Devon, GB) 50°43'N, 3°32'W

ISCA DVMNONIORVM.

Ptol. II,3,13 (Ἴσκα).

It. A. 483.8, 486.8, 486.17.

Tab. Peut.

Rav. Cos. 106.2 (*Scadumnamorum*), 106.6 (*Scadoniorum*).

PNRB 378.

Town on the river Exe, capital of the CIVITAS DVMNONIORVM. A legionary fortress for Legio II Augusta, of 15.4 ha was established c. AD 55 (barracks, baths, *fabrica*); finds suggest evacuation *c.* AD 75. Civil settlement developed *c* AD 80 and continued until the 5th c. Forum and basilica, new baths, walls (late 2nd c) enclosing 37 ha. Port at Topsham, *qv*.

Fox, A., *South-West England,* Newton Abbot, 1973, 166–9: Rainey, 1973, 72: Wacher, 1975, 326–335: Branigan-Fowler, 1976, *passim*: *Britannia* X (1979), 324–6: Webster, 1981, 45–7; Bidwell,

P. T., *The Legionary Bath-house, and Basilica and Forum at Exeter*, Exeter, 1979, and *Roman Exeter, Fortress and Town*, Exeter, 1980.

EXMES Lug., *Esuvii* or *Sagii* (Orne, F) 48°45'N, 0°11'E

'Chaffour et La Briquetière', vicus or minor settlement; cemetery; a well full of deer heads; tegulae, querns, slag.

MSAN VI (1831–33), 437–40; IX (1835), 554–67: *BSHAO* III (1884), 239.

FALAISE Lug., *Viducasses* or *Sagii* (Calvados, F) 48°54'N, 0°12'W

Villa with hypocaust; 18 rooms.

de Caumont, 1838, 153–55; 1850, 453–54: *BSAN* XLIX (1942–45), 99–137, 345.

FALVY Bel., *Viromandui* (Somme, F) 49°48'N, 2°57'E

Villa, 'L'Épine de la Justice'; coin hoard about AD 337.

BSAP X, 41; XI, 196: Vasselle, 314: Agache-Bréart, 1975, 65.

*FAMECHON Bel., *Ambiani* (Somme, F) 49°45'N, 2°03'E

Villa.

Ponchon, 1913: Vasselle, 314: Agache-Bréart, 1975, 65.

FANVM MARTIS *cf* ?Corseul

FARLEY HEATH Brit., *Regni* or *Atrebates* (Surrey, GB) 51°12'N, 0°30'W

Romano-Celtic temple. Temenos, paving, tiles, plaster, votive objects, many coins. Finds suggest that this temple succeeded a pre-Roman shrine and was active until the end of the Romano-British period. Pottery kiln near by, active 4th c.

VCH Surrey IV (1912), 356: *Ant J* XVIII (1938), 391–96: *JRS* XXX (1940), 180; XXXIV (1944), 84: Ross, 1967, 48 etc.

FARMINGTON Brit., *Dobunni* (Gloucestershire, GB) 51°50'N, 1°48'W

Villa, winged corridor; baths. Occupied early 4th c.

TBGAS LXXXVIII (1969), 34–67: McWhirr, 1981, 88.

FARNHAM Brit., *Regni* or *Atrebates* (Surrey, GB) 51°13'N, 0°46'W

Villa, detached bath-building, aqueduct. Five pottery kilns near by. Occupied 3rd and 4th c.

Ant J VIII (1928), 48–50: Lowther, A. W. G., *A Survey of the Prehistory of the Farnham District*, Guildford, 1939, 223–52: *SyAC* LIV (1953–54), 47–57.

FARNINGHAM Brit., *Cantiaci* (Kent, GB) 51°23'N, 0°13'E

Villa, corridor; bath, frescoes, tesserae.

Arch Cant LXI (1948), 180: *JRS* XXXIX (1949), 110: Rivet, 1969, 134, 150.

FAUQUEMBERGUES Bel., *Morini* (Pas-de-Calais, F) 50°36'N, 2°05'E

Gallo-Roman foundations, small finds.

BSAAM XX (1966), 449–67: Leduque, *Morinie*, 65: Delmaire, 1976, 329.

FAVERSHAM Brit., *Cantiaci* (Kent, GB) 51°19'N, 0°54'E

Villa, corridor; frescoes, hypocaust, mosaic.

Philp, B., *Excavations at Faversham, 1965*, Kent Archaeological Research Group, 1968, 62–85.

FÉCAMP Lug., *Caletes* (Seine-Maritime, F) 49°46'N, 0°21'E

Gallo-Roman foundations at 'la Vicomté': cemeteries at 'Val-aux-Vaches' and near the church of Saint-Léonard and at 'La Queue du Renard' (2nd–3rd c). Museum. See also Toussaint ('Camp du Canada').

Cochet, *S.I.*, 370: *BCASM* II, 330: Fallue, L., *Histoire de la ville . . . de Fécamp*, Rouen, 1841.

FEINS Lug., *Redones* (Ille-et-Vilaine, F) 48°20'N, 1°38'W

Gallo-Roman foundations, bricks.

BSAIV LIV (1927–9), 16.

FERMANVILLE Lug., *Unelli* (Manche, F) 49°41'N, 1°27'W

'Hameau du Perray', *vicus* or small settlement; foundations, a kiln, querns, tiles; hoard of 100 large brass coins from Hadrian to Commodus and Crispinus; 2nd c.

MSAN V (1829–30), XLI: de Gerville, 1854, 115: *BSAIC* XXIV (1900), 30: *BSAN* LIII (1955–6), 197–263.

FESQUES Lug., *Caletes* (Seine-Maritime, F) 49°48'N, 1°27'E

Gallo-Roman foundations at 'La Vieux Ville'.

MSAN XI, 174; XXV, 535.

FIERVILLE Lug., *Viducasses* (Calvados, F) 49°04'N, 0°11'W

Small villa with atrium to the SE and SW of the church; painted wall-plaster.

de Caumont, 1831, 227; 1846, 131.

FIFEHEAD NEVILLE Brit., *Durotriges* (Dorset, GB) 50°54'N, 2°19'W

Villa; mosaics, baths, tiles, painted plaster, window glass, jewellery; occupied 3rd and 4th c.

RCHM Dorset III part 1 (1970), 93.

*FIGNIÈRES Bel., *Ambiani* (Somme, F) 49°41'N, 2°35'E

Villa, corridor.

Agache, *BSAP* 1972, 322: Agache-Bréart, 1975, 65: Agache, 1978, 287, 293, 330, 373.

FINCHAMSTEAD Brit., *Atrebates* (Berkshire, GB) 51°22'N, 0°53'W

Milestone on the road between London and Silchester, uninscribed.

VCH Berkshire I (1906), 207: Sedgley, 1975, 18–19 (*sv* Bannisters).

FINCHINGFIELD Brit., *Trinovantes* (Essex, GB) 51°57'N, 0°25'E

Villa, corridor; frescoes.

Rivet, 1969, 144: *VCH Essex* III (1963), 129–30.

FINDON *cf* Muntham Court

FINGRINGHOE WICK Brit., *Trinovantes* (Essex, GB) 51°50′N, 0°58′E
Fort and naval station of Aulus Plautius, AD 43. Destroyed by a sand pit.
Britannia I (1970), 181: Hawkes-Hull, 1947, 19: *VCH Essex* III (1963), 4.

FISHBOURNE Brit., *Regni* (West Sussex, GB) 50°50′N, 0°48′W
Military base, villa and 'palace'. Military base *c* AD 43–44; villa time of Nero; 'palace' built AD 75, burned *c* 280. Perhaps the residence of COGIDVBNVS, client king. Many small finds, fine mosaics. Museum.
Cunliffe, B., *Excavations at Fishbourne*, London, 1971: Rainey, 1973, 73–6: for Cogidubnus, *cf Britannia* X (1979), 227–54.

FITZWORTH HEATH Brit., *Durotriges* (Dorset, GB) 50°41′N, 2°01′W
Salt working, 1st to 2nd c.
RCHM *Dorset* II part 3 (1970), 597.

FIXTINNVM *cf* Meaux

FLÉCHY Bel., *Bellovaci*? (Oise, F) 49°39′N, 2°14′E
Gallo-Roman foundations, tiles.
Leduque, *Ambianie*, 32, n. 130: Agache-Bréart, 1975, 146.

FLEET MARSTON Brit., *Catuvellauni* (Buckinghamshire, GB) 51°50′N, 0°52′W
Minor settlement, on Akeman Street between St Albans and Alchester. Pottery and coins 1st to 4th c., scattered building material, many small finds.
Ordnance Survey Record Card SP 71 NE 4.

*****FLESSELLES** Bel., *Ambiani* (Somme, F) 50°00′N, 2°15′E
Gallo-Roman foundations at 'Longs Champs de Savières'.
Agache, fig. 611: Leduque, *Ambianie*, 129: Agache-Bréart, 1975, 66.

FLEURY/ANDELLE *cf* ?*Ritumagus*

*****FLIXECOURT** Bel., *Ambiani* (Somme, F) 50°01′N, 2°05′E
Cemetery, small finds, weapons, coins.
Leduque, *Ambianie*, 77, n 357: Agache-Bréart, 1975, 67: Agache, 1978, 396, 398.

FOEIL (Le) Lug., *Osismii* (Côtes du Nord, F) 48°25′N, 2°55′W
'la Brousse-Joly', very extensive Gallo-Roman foundations, over more than one hectare; quern.
MSCN I (1883–84), 258.

FOLGOËT (Le) Lug., *Osismii* (Finistère, F) 48°34′N, 4°20′W
'Kergolestroc', villa near the road between Kérilien and l'Aberwrac'h.
BSAF, 1875, 128.

FOLKESTONE Brit., *Cantiaci* (Kent, GB) 51°05′N, 1°12′E
Villa, corridor; bath, hypocaust, mosaics.
Winbolt, S. E., *Roman Folkestone*, London, 1925: *VCH Kent* III (1932), 114: Rivet, 1969, 57–64.

*****FOLLEVILLE** Bel., *Ambiani* (Somme, F) 49°41′N, 2°22′E
Fortress, 'Blamont'.
Agache, fig. 75–77; 1964, 186–89, 387, 388: *Celticum* XV (1966), 139–48: *Gallia* XXI (1963), 342: Agache-Bréart, 1975, 67–68: Agache, 1978, *passim*.

FONGUEUSEMARE Lug., *Caletes* (Seine-Maritime, F) 49°40′N, 0°18′E
Gallo-Roman foundations.
Cochet, *S.I.*, 349.

FONTAINEBLEAU Lug., *Senones* (Seine-et-Marne, F) 48°24′N, 2°42′E
Temple, sanctuaries in the forest of Fontainebleau at 'La Fontaine Sanguinède, La Fontaine St-Aubin'.
Dimier, L., *Fontainebleau* . . ., Paris, 1908: Duval, 1961, 224: Grenier, *Manuel* IV.2 (1960), 716.

FONTAINE-ÉTOUPEFOUR Lug., *Viducasses* (Calvados, F) 49°08′N, 0°27′W
'la Chasse', small villa between Bayeux and Vieux. Many small finds, a dozen coins from the 2nd to the beginning of the 5th c.
de Caumont, 1846, 112: Bertin, Dominique, publication forthcoming.

FONTAINE-SUR-SOMME Bel., *Ambiani* (Somme, F) 50°02′N, 1°56′E
Temple, 'Le Camp Rouge'.
Agache, fig. 88, 108; 1972, 322: Agache-Bréart, 1975, 68: Agache, 1978, 56, 129, 150, 151, 180, 395, 396.

FOOTLANDS Brit., *Regni* (East Sussex, GB) 50°57′N, 0°30′E
Iron workings; furnaces, pottery, terra sigillata 1st–4th c.
Straker, 1931, 327.

FORDWICH Brit., *Cantiaci* (Kent, GB) 51°18′N, 1°07′E
Port for *Durovernum Cantiacorum* (Canterbury). Quay, remains of buildings, all destroyed by a sand pit.
Arch Cant LXII (1949), 145: Rivet, 1964, 145.

FORGES Lug., *Caletes* (Seine-Maritime, F) 49°37′N, 1°33′E
Iron working; minor settlement; slag heaps.
Cochet, *S.I.*, 566.

FORGES-LES-BAINS Lug., *Parisii* (Essonne, F) 48°37′N, 2°06′E
Gallo-Roman foundations, 3rd-c coin hoard.
RN, 6 ser, IX (1967), 140–65: *Gallia* XXV (1967), 213: Roblin, 1971, 189.

FORMERIE Bel., *Ambiani* (Oise, F) 49°39′N, 1°44′E
Gallo-Roman foundations, small finds.
Leduque, *Ambianie*, 33, n 138.

FOUCARMONT Lug., *Caletes* (Seine-Maritime, F) 49°51′N, 1°34′E
Gallo-Roman foundations in the abbey grounds, 'le Catelier, le Camp du Bourg, le Fond Théodore and l'Épinette'.
Cochet, *S.I.*, 550.

FOUCART Lug., *Caletes* (Seine-Maritime, F) 49°37′N, 0°36′E
Minor settlement.
BSNÉP XXXVI, fasc. 1; XXXVI, fasc. 4; XXXVIII, fasc. 2.

*FOUENCAMPS Bel., *Ambiani* (Somme, F) 49°50′N, 2°25′E
Gallo-Roman foundations, ponds, hypocaust (villa).
BSAP VIII, 75: Vasselle, 316: Agache-Bréart, 1975, 69: Agache, 1978, 73, 253, 262, 267.

FOULNESS Brit., *Trinovantes* (Essex, GB) 51°34′N, 0°51′E
Tumulus; large urn, pottery, terra sigillata.
VCH Essex III (1963), 132.

FOXCOMBE HILL Brit., ?*Dobunni* (Oxfordshire, GB) 51°42′N, 1°17′W
Pottery kilns, 1st to 4th c.
Oxon XIII (1948), 32–38; XVII–XVIII (1952–53), 229–31.

*FRAMERVILLE Bel., *Ambiani* (Somme, F) 49°52′N, 2°42′E
Large villa, quarries, sarcophagi.
Agache, fig. 308, 448: Leduque, *Ambianie*, vii: Agache-Bréart, 1975, 69–70.

FRAMPTON Brit., *Durotriges* (Dorset, GB) 50°45′N, 2°33′W
Villa; mosaics, with chi-ro.
RCHM *Dorset* I (1952), 150: *PDNHAS* LXXVIII (1956), 81–3: Rainey, 1973, 77–9.

FRANCASTEL Bel., *Bellovaci* (Oise, F) 49°35′N, 2°09′E
Gallo-Roman foundations, many sarcophagi, pottery, coins.
Leduque, *Ambianie*, 32, n 132: Agache, 1978, 351.

FRANCIÈRES Bel., *Ambiani* (Somme, F) 50°04′N, 1°56′E
Gallo-Roman foundations; coin hoard, Trajan-Hadrian.
Agache, fig. 643: Leduque, *Ambianie*, 171, n 826: Agache-Bréart, 1975, 70: Agache, 1978, 215, 448.

FRANQUEVILLETTE Lug., *Veliocasses* (Seine-Maritime, F) 49°25′N, 1°15′E
Villa, gallery façade, at 'Les Longues pièces'; hypocaust, plaster, terra sigillata, end of 2nd c?
Vesly, L., *BCASI* XV (1909), 113, 117–19; *Exploration du plateau de Boos,* notes, 1910.

FRÉMÉCOURT Lug., *Veliocasses* (Val d'Oise, F) 49°07′N, 2°00′E
Villa.
Gallia XIX (1961), 291.

FRÉNOUIVILLE Lug., *Viducasses* (Calvados, F) 49°08′N, 0°14′W
CIL XIII 8990.
Villa on the road between Vieux and Lisieux; milestone of Trajan. Coin hoard from Severus to Maximinus and 25 coins of 1st to 4th c. Some 800 Gallo-Roman and Merovingian burials; a slab of chalk bearing a cross pattée.
de Caumont, 1850, 119: *MSAN* XXVIII (1870),

74–75: *BSAN* XI (1883), 612–14; LI (1948–51), 357: *Gallia* XXVI (1968), 347; XXVIII (1970), 271–72; XXX (1972), 336–38.

FRESNES-LÈS-MONTAUBAN Bel., *Atrebates* (Pas-de-Calais, F) 50°20′N, 2°56′E
Villa, cemetery.
Jelski, 140: Agache-Bréart, 1975, 138.

*FRESNES-MAZANCOURT Bel., *Ambiani* (Somme, F) 49°51′N, 2°52′E
Minor settlement, 'Genermont, les Tombes à Moyenneville'; coins.
BSAP IX, 199: Agache, fig. 584: Agache-Bréart, 1975, 70: Agache, 1978, 215, 422.

*FRESNOY-EN-CHAUSSÉE Bel., *Ambiani* (Somme, F) 49°46′N, 2°35′E
?SEEVIAE (*qv*).
Major settlement at St Mard.
MSAP I (1838), 476: Vasselle, 316: Agache, fig. 553: Leduque, *Ambianie*, 142, n 282: Agache-Bréart, 1975, 70: Agache, 1978, 423.

FRESNOY-FOLNY Lug., *Caletes* (Seine-Maritime, F) 49°53′N, 1°26′E
Villa, beside the road between Eu and Neufchâtel.
BCASM, 1867, 23, 77: Cochet, *S.I.*, 538.

*FRESNOY-LÈS-ROYE Bel., *Ambiani* (Somme, F) 49°44′N, 2°47′E
Villa; coin hoard.
Agache, fig. 290: *BSFN*, July 1968, 288–89: *MSAP* XXIII (1971): Agache-Bréart, 1975, 71.

FRESQUIENNES Lug., *Caletes* (Seine-Maritime, F) 49°34′N, 1°00′E
Cemetery, end of 1st c BC, at the hamlet of Caillotère.
BCASM XXI, 117.

FRETTEMEULE Bel., *Ambiani* (Somme, F) 50°01′N, 1°39′E
Gallo-Roman foundations, 'Maigneville'.
Vasselle, 316.

FRETVM GALLICVM
The Dover Strait, between the North Sea and the English Channel (*Oceanus Germanicus* and *Oceanus Britannicus*).
Tac. *Agric.* 40 (*fretum Oceani*).
Solinus XXII, 8.
Isidorus XIV, 6.

FRILFORD Brit., *Atrebates*/*Dobunni* (Oxfordshire, GB) 51°40′N, 1°22′W
Romano-Celtic temple on site of British temple? 2nd c BC. A circular stone structure replacing the original was itself replaced by a rectangular building with temenos, ? c 4th c AD. Many votive objects, amphitheatre nearby.
Oxon IV (1939), 1–70: Ross, 1967, 45: *Britannia* XIII (1982), 305–9.

FRILFORD Brit., *Atrebates*/*Dobunni* (Oxfordshire, GB) 51°40′N, 1°23′W
Villa, corridor; detached bath-house, hypocausts. Coins of 4th c.
VCH Berkshire I (1906), 207–09.

FROCESTER COURT Brit., *Dobunni*

(Gloucestershire, GB) 51°43′N, 2°19′W

Villa; baths, mosaics, pottery, coins, jewellery, bones. Occupied c AD 275 to early 5th c.

TBGAS LXXXIX (1970), 15–86; Rainey, 1973, 79: McWhirr, 1981, 83–7.

FULLERTON Brit., *Belgae* (Hampshire, GB) 51°09′N, 1°28′W

Villa, corridor; one mile south of Wherwell; mosaics.

Athenaeum, 25 Feb. 1905, 250: *JRS* XII (1922), 250; LIV (1964), 17; LV (1965), 217: Rainey, 1973, 79.

FULMER Brit., *Catuvellauni* (Buckinghamshire, GB) 51°34′N, 0°34′W

Pottery kiln, 2nd c AD.

Records of Buckinghamshire XIV (1941–46), 67–69, 153–63.

FYFIELD *cf* Lambourne's Hill

GADANCOURT *cf* Guiry-en-Vexin

GADEBRIDGE PARK Brit., *Catuvellauni* (Hertfordshire, GB) 51°46′N, 0°28′W

Villa. Timber house AD 175, stone winged-corridor built 2nd c. Enlarged c 200 and again c 300. Destroyed mid-4th c.

Neal, D. S., *The Excavation of the Roman Villa in Gadebridge Park, Hemel Hempstead, 1963–68,* London, 1974: Todd, 1978, 33–58, *passim.*

GAILLON *cf* Condécourt

GALLEY HILL Brit., *Cantiaci* (Kent, GB) 51°27′N, 0°17′E

Pottery kiln.

Arch Cant XXVII (1905), lxxiii–lxxiv: *VCH Kent* III (1932), 131.

GAMACHES Bel., *Ambiani* (Somme, F) 49°59′N, 1°33′E

Villa; walls, frescoes, tiles, marble, coins, hypocaust.

Leduque, *Ambianie,* 168: Vasselle, 318: Agache, 1978, 40, 41, 253, 262, 267.

GATCOMBE Brit., ?*Belgae* (Avon, GB) 51°25′N, 2°41′W

Villa within walls built 4th c and enclosing about 6.5 ha, and other buildings including a 3rd-c basilica. Evidence of iron- and lead-working. Occupied from 1st to 5th c.

PSANHS CXI (1966–67), 24–35; CXII (1967–68), 40–43: *PUBSS* XI/2 (1967), 126–60: Rodwell-Rowley, 1975, 175–82, 114.

GAUBERTIN Lug., *Carnutes* (Loiret, F) 48°07′N, 2°25′E

Gallo-Roman foundations, cemetery; at Entraigues', statuette of Epona.

Nouel, 21, 37, 38.

GELLIGAER Brit., *Silures* (Mid Glamorgan, GB) 51°40′N, 3°15′W

RIB 397–400.

Fort and practice camps. Flavian fort with earthen rampart; Trajanic stone fort with stone-founded internal buildings. Five practice camps on G. Common nearby to northwest.

Nash-Williams, 1969, 88–91: *JRS* LIX (1969), 126 RCAHM (Wales), *Glamorgan* I, ii (1976), 95–8, 103–5.

GENAINVILLE Lug., *Veliocasses* (Val d'Oise, F) 49°08′N, 1°45′E

Temple at 'Château-Bicêtre, Les Vaux de la Celle'; central building with double cella (?Mercury) in a gallery with niches, nymphaeum, dwellings, sculpture, frescoes, 2nd half of 2nd c; theatre.

Duval, 1961, 237: *Gallia* VII (1949), 111; XIX (1961), 293; XXI (1963), 349; XXIII (1965), 304; XXV (1967), 219; XXVIII (1970), 246; XXX (1972), 309.

GESNES-LE-GANDELIN Lug., *Aulerci Cenomani* (Sarthe, F) 48°21′N, 0°01′E

Fort?, on a rise 2 km distant, 'camp de St-Erroul, or Entre-Vaux'; 250×140 m, ramparts of earth and stone 5 to 12 m high; to the east a gully, one entrance on the west, 1 km from a Roman road.

Bull.Soc.sc.et arts Sarthe V, 202–06; VIII, 152: Ledru, 1911, 91: *RHAM* XXXV, 121: Voisin, 1862, 76.

GESOCRIBATE *cf* ?Brest

GESORIACVM *cf* Boulogne-sur-Mer

GIBERVILLE Lug., *Viducasses* (Calvados, F) 49°11′N, 0°17′W

Gallo-Roman foundations; seven square paved pits, possibly a tannery. A Gallo-Roman sarcophagus; a bronze tripod. Coins of Antoninus Pius, Aelius, Crispinus, Septimius Severus and Postumus; 2nd and 3rd c.

de Caumont, 1850, 33.

GIEL Lug., *Esuvii* or *Sagii* (Orne, F) 48°46′N, 0°12′W

Small villa.

MSAN IX (1835), 461.

GIROLLES Lug., *Senones* (Loiret, F) 48°04′N, 2°43′E

?VELLAVNODVNVM (*qv*).

Gallo-Roman foundations at 'Grandvillon'.

Nouel, 21, 41.

GISACVM

CIL XIII.1, 3197 (*Deo Gisaco*); 3204 (*Gisaci civis*); 3488 (*Gesaco*).

Name derived from deity. Site identified with Gisay-la-Coudre or Vieil-Évreux (*qqv*).

GISAY-LA-COUDRE Lug., *Aulerci Eburovices* (Eure, F) 48°57′N, 0°38′E

?GISACVM (*qv*).

Gallo-Roman foundations.

GLACERIE (La) Lug., *Unelli* (Manche, F) 49°36′N, 1°34′W

Gallo-Roman foundations, ?fort, at 'Grandcamp', 'les Buissonets'; coins.

BSAIC XXIV (1900), 59–60.

GLANVILLE Lug., *Lexovii* (Calvados, F) 49°17′N, 0°04′E

53

Temple constructed of flints and shore pebbles bonded with mortar; roof-tiles, iron nails. An inhumation in the gallery.

MSAN XII (1841), 428: de Caumont, 1862, 65: Vesly, 13–14.

GLEVENSES *cf* Gloucester

GLEVVM COLONIA *cf* Gloucester

*GLISY Bel., *Ambiani* (Somme, F) 49°52′N, 2°24′E

Small villa, many coins.

Agache, fig. 34, 274: Leduque, *Ambianie*, 137, n 201: Agache-Bréart, 1975, 73.

GLOUCESTER Brit., *Glevenses* (Gloucestershire, GB) 51°52′N, 2°14′W

GLEVVM COLONIA.

It. A. 485.4 (*Clevo*).

Rav. Cos. 106.29 (*Glebon Colonia*).

CIL VI, 3346 (*Ner Glevi*); XVI, 130 (*Glevi*).

RIB 119–24, 161 (*Coloniae Glev*).

JRS XIX (1929), 216, 8; XLVII (1957), 233, 31; LI (1961), 196, 32; LII (1962), 197, 40.

PNRB 368–9.

Town at the lowest crossing of the river Severn, in territory taken from the *Dobunni*. Fort (unlocated) of *Cohors VI Thracum Equitata c* AD 43–9 (*RIB* 121); at Kingsholm (1 km N of the later town centre) perhaps fortress for part of *Legio XX Valeria Victrix* 49–55; at G. itself fortress perhaps of *Legio XIV Gemina Martia Victrix* 69–70, then of *Legio II Augusta* until 75. Colonia founded by Nerva *c* 97, officially named COLONIA NERVIA GLEVENSIVM. Forum and basilica, possible baths, timber houses later rebuilt in stone. Walls (rebuilt *c* AD 200) enclosing 18.4 ha, with large extra-mural settlement to N and W. Mosaics, pottery, coins, many small finds. Museum. G. was taken by the Saxons after the battle of Dyrham, AD 577.

JRS XXXII (1942), 39–52: Ross, 1967, 88 etc.: Rainey, 1973, 81–5: *Britannia* IV (1973), 309; VI (1975), 273; VII (1976), 354; X (1979), 319–22; XI (1980), 73–114: *Ant J* LII (1972), 24–69, LIV (1974), 8–52: Wacher, 1975, 137–55: Branigan-Fowler, 1976, 63–80: McWhirr, 1981, 21–58: Webster, 1981, 43–4.

GOBANNIVM *cf* Abergavenny

GOLDHANGER Brit., *Trinovantes* (Essex, GB) 51°44′N, 0°45′E

Salt workings; briquetage, pottery, terra sigillata, evaporation vessels, 1st c.

Nenquin, 1961, 83: *VCH Essex* III (1963), 32–34, 134.

GONFREVILLE-L'ORCHER Lug., *Caletes* (Seine-Maritime, F) 49°30′N, 0°16′E

Gallo-Roman foundations, traces of ? villa under the chapel of Saint-Dignefort.

MSAN XII (1841), 177.

GORHAMBURY Brit., *Catuvellauni* (Hertfordshire, GB) 51°46′N, 0°23′W

Villa, winged corridor. Circular late Iron Age dwellings succeeded by timber house burnt in 1st c; early 2nd c masonry house, later extended and altered in 2nd c and 4th c. Baths, hypocausts, cellar.

TStAHAAS, 1961, 21–30: *Britannia* IV (1973), 299; V (1974), 437; VI (1975), 258; VII (1976), 339–40; VIII (1977), 401–2; IX (1978), 445; X (1979), 305–6; XI (1980), 373–4: Todd, 1978, 33–58, *passim*.

GOSBECK'S FARM Brit., *Trinovantes* (Essex, GB) 51°51′N, 0°51′E

Temple, Romano-Celtic, square on plan within its temenos, on the site of a prehistoric temple; end *c* AD 400. Isolated theatre; traces of habitations, roads, on aerial photograph; lime kiln. Claudian or Neronian fort nearby. *Cf* Stanway.

Hull, 1958, 259–71: *JRS* LVIII (1968), 196: Lewis, 1966, *passim*: Ross, 1967, 51: Dunnett, 1975, 114–8: *Britannia* II (1971), 27–47, VIII (1977), 88, 185–7.

GOUAREC Lug., *Osismii* (Côtes du Nord, F) 48°13′N, 3°11′W

Gallo-Roman foundations in the bed of the Blavet; two white stone statuettes; many large bronze coins from Augustus to Gordian.

MSCN I (1883–84), 502.

GOUÉZEC Lug., *Osismii* (Finistère, F) 48°10′N, 3°58′W

'Moguerou' = 'les murailles', villa, probable temple, near the road between Quimper and Morlaix. Building debris; ancient mines (?iron) and metallurgical industries in the vicinity, not dated.

BSAF, 1968, 7.

GOUILLONS Lug., *Carnutes* (Eure-et-Loir, F) 48°21′N, 1°50′E

Villa, 'Les Hauts de Châtenay'; coin hoard.

Air photograph, Jalmain: Nouel, 36.

GOULET Lug., *Esuvii* or *Sagii* (Orne, F) 48°44′N, 0°05′W

'Camp de César', possible fort. Ramparts 3,60 m high with external ditch; seven wells.

MSAN IX (1835), 444.

GOUVIEUX Bel., *Silvanectes* (Oise, F) 49°11′N, 2°25′E

Gallo-Roman foundations; oppidum, 'Le Camp de César'.

Bull Soc arch Creil XCVIII (1977), 15–20; XCIX (1978), 4–15: Roblin, 1978, 27 n 14.

*GOUY-L'HÔPITAL Bel., *Ambiani* (Somme, F) 49°50′N, 1°57′E

Gallo-Roman foundations.

Leduque, *Ambianie*, 171, n 816: Agache, 1978, 215.

GOUY-SERVINS Bel., *Atrebates* (Pas-de-Calais, F) 50°24′N, 2°39′E

Gallo-Roman foundations, square in plan.

Jelski, 143.

GRAIN, Isle of Brit., *Cantiaci* (Kent, GB) 51°27′N, 0°42′E

Pottery kiln.

Arch Cant LXII (1949), xlv.

GRAINCOURT Lug., *Caletes* (Seine-Maritime,

F) 49°56′N, 1°10′E
Minor settlement extending equally over Derchigny and Bracquemont.
Cochet, *S.I.,* 254.

GRAINVILLE-L'ALOUETTE Lug., *Caletes* (Seine-Maritime, F) 49°39′N, 0°27′E
Cremation cemetery at 'le Fief des Champs', later 'la Terre à Pots'.
Cochet, *S.I.,* 380.

GRAINVILLE-LA-TEINTURIÈRE Lug., *Caletes* (Seine-Maritime, F) 49°45′N, 0°38′E
?GRAVINVM (*qv*).
Tab. Peut.
?GRANNONA (*qv*).
Not. Dig. Occ. XXXVII, 14 and 23.
Major settlement. Traces of buildings; tradition of a lost town. Coins, tombs.
Cochet, *S.I.,* 453: Desjardins, *Géo. Gaule,* 293, 330 ff.

GRAND-COURONNE Lug., *Veliocasses* (Seine-Maritime, F) 49°22′N, 1°00′E
Villa with hypocaust at 'Grésil'; temple at 'Essarts'. *Cf* Rouvray.
Bull arch, 1903, 103; 1905, 66: Vesly, 84: *Gallia* XXX (1972), 346.

GRANDCOURT Lug., *Caletes* (Seine-Maritime, F) 49°55′N, 1°30′E
Gallo-Roman foundations, pottery widely scattered.
Cochet, *S.I.,* 538.

GRANDES-VENTES (Les) Lug., *Caletes* (Seine-Maritime, F) 49°49′N, 1°14′E
Gallo-Roman foundations at the hamlet of Orival; at that of Le Châtelet, tradition of a vanished town called Hesdin.
Cochet, *S.I.,* 272.

GRANNONA
Not. Dig. Occ. XXXVII, 14 and 23.
Fort of the *Tractus Armoricanus et Nervicanus,* 4th c. Identification disputed: Banville? Grainville-la-Teinturière? Le Havre?
Johnson, 1979, 74, 90, 93–4.

GRATELEY Brit., *Belgae* (Hampshire, GB) 51°10′N, 1°36′W
Villa, corridor. Foundations, mosaics, wall plaster, coin of Gallienus.
PHFC VI (1907–10), 341–42; VIII (1917), 107–08: Rainey, 1973, 85.

GRATTEPANCHE Bel., *Ambiani* (Somme, F) 49°47′N, 2°17′E
GRATIANI PAGVS.
Cemetery, 'Le Mont de César'; funerary pits, coins of Antoninus Pius.
Leduque, *Ambianie,* 149, n 442–43.

GRAVINVM.
Tab. Peut.
Disputed locality: towards Fécamp or near Lillebonne, Gonfreville-l'Orcher or Grainville-la-Teinturière (*qv*)?

GREAT BRAXTED Brit., *Trinovantes* (Essex, GB) 51°47′N, 0°42′E
Tile kiln.
VCH Essex III (1963), 57.

GREAT DUNMOW Brit., *Trinovantes* (Essex, GB) 51°52′N, 0°21′E
Major settlement on the road between Braughing and Colchester. Foundations, cemetery.
Rivet, 1964, 147: *VCH Essex* III (1963), 125: Rodwell-Rowley, 1975, 85–101.

GREAT TEW Brit., *Dobunni* (Oxfordshire, GB) 51°56′N, 1°25′W
Villa, courtyard. Baths, mosaics.
VCH Oxfordshire I (1939), 310–11: Rainey, 1973, 89.

GREAT WITCOMBE Brit., *Dobunni* (Gloucestershire, GB) 51°49′N, 2°09′W
Villa, courtyard. Built late 1st c, occupied into the 5th. Baths, mosaics, coins of 4th c.
TBGAS LXXIII (1954), 5–69: *Britannia* I (1970), 295: Rainey, 1973, 89: McWhirr, 1981, 92–3.

GREAT WYMONDLEY Brit., *Catuvellauni* (Hertfordshire, GB) 51°57′N, 0°15′W
Villa; hypocausts, mosaics, tiles, coins of 3rd and 4th c, many small objects.
VCH Hertfordshire IV (1914), 170–71: Rainey, 1973, 90.

GREEN ORE Brit., ?*Belgae* (Somerset, GB) 51°16′N, 2°37′W
Lead workings; pottery, pigs of lead stamped IMP. VESPASIAN. AVG. (AD 69–79).
PSANHS CI (1956–57), 52–88.

GREENWICH PARK Brit., ?*Londinienses* (Greater London, GB) 51°28′N, 0°00′
Romano-Celtic temple; statue (feminine) 2/3 natural size; coins, Claudius to Honorius, inscriptions, tesserae, remains of walls, columns. Museum.
VCH Kent III (1932), 16: *RIB* 14: Lewis, 1966, 126, 141: *LA* III.12 (1979), 311–317.

GRÉGY-SUR-YERRES Lug., *Parisii* (Seine-et-Marne, F) 48°40′N, 2°37′E
Villa.
Duval, 1961, 230, 234: Roblin, 1971, 56, 158, 322.

GRISELLES Lug., *Senones* (Loiret, F) 48°05′N, 2°50′E
Cemetery.
Nouel, 27.

***GRIVESNES** Bel., *Ambiani* (Somme, F) 49°41′N, 2°29′E
Temple, 'Cimitière St-Agnan'; Gallo-Roman foundations, ?vicus.
BSAP XXIV, 403: Agache, fig. 220, 243; 1972, 322: Vasselle, 318: Agache-Bréart, 1975, 73–4.

GROSVILLE Lug., *Unelli* (Manche, F) 49°31′N, 1°44′W
Aqueduct connected with the river Scie.
BSAIC XXIV (1900), 71.

GUÉMY Bel., *Morini* (Pas-de-Calais, F) 50°48′N, 2°02′E

Cremation cemetery. ?fort.
Ringot, 176.

GUERN Lug., *Veneti* (Morbihan, F) 48°02′N, 3°05′W
At 'Petit Bonalo' cemetery, bronzes. At 'Pradigo', pottery and great quantities of bricks.
BSPM, 1902, 46–47; 1931, 17.

GUERNSEY ?Lug., *?Unelli* Channel Islands (GB) ÎLE DE GUERNESEY Les Îles Anglo-Normandes 49°28′N, 2°34′W
?LISIA INSVLA.
It. M. 509.2 (*Silia* or *Lisia*).
PNRB 181.

GUILER-SUR-GOYEN Lug., *Osismii* (Finistère, F) 48°01′N, 4°22′W
'Lansaludou', Gallo-Roman foundations of what might be a villa; hypocausts.
BSAF, 1972, 66.

GUILERS Lug., *Osismii* (Finistère, F) 48°26′N, 4°34′W
'Kerroual', Gallo-Roman foundations of what might be a villa. Aureus of Nero, sestertius of Antoninus Pius.
BSAF, 1875, 129: du Chatellier, 1907, 111.

*****GUILLAUCOURT** Bel., *Ambiani* (Somme, F) 49°51′N, 2°38′E
Temple with important associated structures, 'Fond d'Enguillaucourt'.
Agache, fig. 572; *BSAP,* 1972, 322: Agache-Bréart, 1975, 74: Agache, 1978, 260, 392, 396, 398, 402, 438.

GUILLEVILLE Lug., *Carnutes* (Eure-et-Loir, F) 48°13′N, 1°49′E
Gallo-Roman foundations, statuary.
Nouel, 20, 29.

GUIMILIAU Lug., *Osismii* (Finistère, F) 48°29′N, 4°00′W
'Creac'h ar Bléis', Gallo-Roman foundations. Two aurei of Antoninus Pius.
BSAF, 1875, 129.

GUIPAVAS Lug., *Osismii* (Finistère, F) 48°26′N, 4°24′W
'Cosquerou', Gallo-Roman foundations of what might be a villa near the road between Landerneau and St Reman and the sea. Fragments of metal; coarse pottery, amphorae, terra sigillata IVLLVS. MONTANS. Nero and Trajan.
Not published.

GUIRY-EN-VEXIN and **GADANCOURT** Lug., *Veliocasses* (Val d'Oise, F) 49°08′N, 1°51′E
?PETROMANTALVM.
It. A. 382.5, 384.10.
Tab. Peut.
Villa, 'Les Terres Noires'. One principal building with bath-house; three other buildings, one an outhouse, have been excavated; several others not yet examined in a walled enclosure 200 m long each side. 1st to 4th c; mainly 2nd c, interruption in the 1st half of 3rd c, burned 276. Numerous small finds. Museum.

BAVF I (1965), 65 sq; II (1966), 49 sq; III (1967), 97: *Gallia* XVI (1958), 266 sq; XVII (1959), 274; XVIII (1960), 163; XIX (1961), 291.

GUISSENY Lug., *Osismii* (Finistère, F) 48°38′N, 4°25′W
'le Curnic', a fishpond (vivier) on the beach. Eight fibulae, two couples in bronze. Coins of Augustus, Trajan, Hadrian, Faustina, Commodus, Tetricus. Terra sigillata Dr 35, 46; pottery, glass.
Ann Bret LXXV (1968), 246–65.

GUITTÉ Lug., *Coriosolites* (Côtes du Nord, F) 48°18′N, 2°06′W
'le Clos Long', villa; building debris.
BSCN XLVII (1909), 12–13.

GWENNAP *cf* Mynheer Farm

HADLEIGH Brit., *Trinovantes* (Essex, GB) 51°32′N, 0°36′E
Crop-mark of ditched enclosure, once claimed as military, but more probably a farmstead. *Cf.* Orsett.
JRS XLIII (1953), 97; Dunnett, 1975, 41–2.

HALATTE (Forêt d') Bel., *Silvanectes* (Oise, F) 49°15′N, 2°37′E
Temple, enclosure 80 by 140 m, 'Malgenest'; 2nd c.
Gallia XXV (1967), 197; XXVII (1969), 234: Grenier, *Manuel* IV.2 (1960), 815–6.

HALINGHEN Bel., *Morini* (Pas-de-Calais, F) 50°36′N, 1°42′E
?VICVS DOLVCENSIS (corruption of *Dolichenus*?).
CIL XIII, 3563.
Minor settlement ('Mont Violette', 'Camp de César', 'Le Haut Tingry'). 800 m from the church. Abandoned 4th c for a site in the valley. Bases, tombs, in the foundations of the church.
Leduque, *Boulonnais,* n 29–31; *Morinie,* 62: Delmaire, 1976, 136.

HALLCOURT WOOD Brit., *Belgae* (Hampshire, GB) 50°55′N, 1°13′W
Pottery kilns, late 1st c.
PHFC XXII (1961), 8–21.

HALLOY-LÈS-PERNOIS Bel., *Ambiani* (Somme, F) 50°03′N, 2°12′E
Temple, and ?villa, 'L'Hôtel-Dieu'.
Agache, *BSAP,* 1972, 322: Agache-Bréart, 1975, 76: Agache, 1978, 398.

HALSTEAD Brit., *Trinovantes* (Essex, GB) 51°56′N, 0°38′E
Pottery kiln.
VCH Essex III (1963), 137.

HALSTOCK Brit., *Durotriges* (Dorset, GB) 50°52′N, 2°40′W
Villa; mosaics, baths.
RCHM Dorset I (1952), 121: Rainey, 1973, 91–92.

HAMBLAIN-LES-PRÉS Bel., *Atrebates* (Pas-de-Calais, F) 50°18′N, 2°58′E
Temple: villa.

Dérolez, 519: Jelski, 137.

HAMBLEDEN Brit., *Catuvellauni* (Buckinghamshire, GB) 51°34′N, 0°52′W
Villa; baths, tessellated pavements; occupied mid-1st to late 4th c.
Arch LXXI (1921), 141–98.

HAM HILL Brit., *Durotriges* (Somerset, GB) 50°57′N, 2°45′W
Stone quarry inside an Iron Age hill fort probably originating in late 1st c BC or early 1st c AD. Durotrigian coins and Roman military equipment incl. part of a *lorica squamata*.
VCH Somerset I (1906), 295–98: *Arch J* CVII (1950), 90–91, CXV (1958), 54, 81–3.

HAM HILL Brit., *Durotriges* (Somerset, GB) 50°57′N, 2°44′W
Villa, corridor. Situated within the defences of an Iron Age hill fort. Mosaics, coins mid-3rd to mid-4th c.
JRS III (1913), 127–33.

HAMMER WOOD Brit., *Regni* (East Sussex, GB) 51°05′N, 0°02′E
Iron workings; pottery, cinders.
O.S. fiche TQ 43 NW 5.

HAMSTEAD MARSHALL Brit., *Atrebates* (Berkshire, GB) 51°24′N, 1°24′W
Pottery kiln, 4th c.
ANL VI.12 (1960), 285.

*****HANGARD** Bel., *Ambiani* (Somme, F) 49°49′N, 2°31′E
Villa; coins, urns.
Agache, fig. 479: Leduque, *Ambianie,* 142: Agache-Bréart, 1975, 76: Agache, 1978, 72, 215.

HANNINGTON Brit., ?*Dobunni* (Wiltshire, GB) 51°39′N, 1°44′W
Villa; tessellated pavements, coin of early 4th c.
VCH Wiltshire I (1957), 75.

HANVEC Lug., *Osismii* (Finistère, F) 48°20′N, 4°10′W
'Pointe du Gligeau', villa; abundant terra sigillata, much from the end of 1st c. On an Iron Age site.
BSAF, 1969, 26–31.

HARCOURT Lug., *Lexovii* (Eure, F) 49°10′N, 0°48′E
Villa.
Congrès archéologique de France LVI (Évreux, 1889), 243–9.

*****HARDECOURT-AU-BOIS** Bel., *Ambiani* (Somme, F) 49°59′N, 2°49′E
Temple, 'La Garenne, Les Cerisiers'.
Agache, *BSAP,* 1972, 322: Agache-Bréart, 1975, 77.

HARDINGHEN Bel., *Morini* (Pas-de-Calais, F) 50°48′N, 1°49′E
Cemetery.
Ringot, 178.

HARDHAM Brit., *Regni* (West Sussex, GB) 50°57′N, 0°32′W
Minor settlement. On Stane Street between London and Chichester. Earthen rampart enclosing 1.6 ha. Pottery and other small finds, six ritual shafts. Occupied 1st and 2nd c.
SxAC XVI (1864), 52–64; LXVIII (1927), 89–132: Coles, J., and Simpson, D. D. A., eds., *Studies in Ancient Europe,* Leicester, 1968, 265: Cunliffe, 1973, 69–70.

HARFLEUR Lug., *Caletes* (Seine-Maritime, F) 49°30′N, 0°11′E
CARACOTINVM.
It. A. 381.7 (*Caracotino*).
Major settlement and port, 'Caucriauville', on the estuary of the river Seine. 'Mont-Cabert', settlement of 2nd to 4th c; Gallo-Roman foundations, temple ('Côte des Buquets'), cremation cemetery, figurine, pottery.
MSAN XII (1841), 117: *BCASI* IX.3 (1893), 397: Vesly, 12–18: Deglatigny, 1931, 149: Duval, 1961, 230, 237, 243: *RSSHN* XXX (1963), 7–16: *Gallia* XXII (1964), 292; XXIV (1966), 270; XXVI (1968), 372: *BSNÉP* XXXVIII, II, 79; XXXIX, I, 9.

HARLOW Brit., *Trinovantes* (Essex, GB) 51°46′N, 0°07′E
Temple in its temenos replacing a proto-historic temple; built 1st c, destroyed early in the 5th c; inscriptions.
VCH Essex III (1963), 139: *JRS* LVI (1966), 210: Lewis, 1966, *passim*: Ross, 1967, 50, 69: *C Arch* I (1968), 287–90.

HARMOYE (La) Lug., *Coriosolites* (Côtes du Nord, F) 48°20′N, 2°58′W
'Kerguz', Gallo-Roman foundations near the road between Saint-Brieuc and Corlay.
MSCN I (1883–84), 248.

HARNES Bel., *Atrebates* (Pas-de-Calais, F) 50°27′N, 2°54′E
Gallo-Roman foundations; coins.
Gallia XXIX (1971), 230: Leduque, *Atrébatie,* 99.

HARPENDEN *cf* Pickford Mill

HARPSDEN Brit., *Catuvellauni* (Oxfordshire, GB) 51°31′N, 0°54′W
Villa, aisled, with baths.
VCH Oxfordshire I (1939), 323–24.

HARTLAND POINT Brit., *Dumnonii* (Devon, GB) 51°01′N, 4°31′W
HERCVLIS PROMONTORIVM.
Ptol. II.3.2 ('Ηρακλέους ἄκρον).
PNRB 372.

HARTLIP Brit., *Cantiaci* (Kent, GB) 51°20′N, 0°36′E
Villa, type unknown; large bath building.
VCH Kent III (1932), 117–8.

HASSOCKS Brit., *Regni* (West Sussex, GB) 50°55′N, 0°09′W
Minor settlement. Cemetery in use 1st to 3rd c. Pottery, coins.
SxAC LXVI (1925), 34–61: Cunliffe, 1973, 72 etc.

HASTINGS Brit., *Regni* (East Sussex, GB) 50°51′N, 0°34′E
Port, for the export of iron. Mosaics, cemetery,

coins. Museum.
SxAC XIV (1862), xiii: *VCH Sussex* III (1935), 58: Rivet, 1964, 145.

HAUTOT-SUR-MER Lug., *Caletes* (Seine-Maritime, F) 49°54′N, 1°02°E
Gallo-Roman foundations near the church.
Cochet, *S.I.,* 240.

HAVANT, LANGSTONE Brit., *Regni* (Hampshire, GB) 50°50′N, 0°59′W
Villa, corridor type. Baths, hypocausts; coins Vespasian to Constans.
PHFC X (1930), 286–87.

HAVANT, LITTLEPARK Brit., *Regni* (Hampshire, GB) 50°52′N, 1°01′W
Villa; earlier building *c* AD 150–200; corridor house *c* 250–350. Tiles, painted plaster, pottery.
JRS XVI (1926), 232–33.

HAVRE (Le) Lug., *Caletes* (Seine-Maritime, F) 49°30′N, 0°07′E
?GRANNONA (*qv*).
Gallo-Roman foundations at Sainte-Adresse; cremation cemeteries in Ingouville, Graville, Bois des Halattes, in Rue Montmirail and in the Forêt de Montgeon. Musée de l'Abbaye de Graville, Musée de l'Ancien Havre.
Cochet, *S.I.,* 335–7: *BSNÉP* XXXV, fasc. 4, 114: Johnson, 1979, 93–4.

HAŸ-LES-ROSES Lug., *Parisii* (Val-de-Marne, F) 48°47′N, 2°20′E
Gallo-Roman foundations, extensive; amphora factory 1st to 4th c, tilery, dwelling with columns, lake, votive pits, burials 3rd c, coins. *Cf* Chevilly-Larue.
Duval, 1961, 173, 176, 231: Roblin, 1971, 65, 282.

HAYLING Brit., *Regni* (Hampshire, GB) 50°49′N, 0°58′W
Circular temple within square *temenos,* constructed mid- or late 1st c on site of an Iron Age shrine.
T. Ely, *Roman Hayling* (London, 1904, 1908): *Britannia* VII (1976), 366; VIII (1977), 418; IX (1978), 463–4; X (1979), 329–31; XI (1980), 393; XII (1981), 361, 369.

HEATHFIELD Brit., *Regni* (East Sussex, GB) 50°59′N, 0°18′E
Iron workings; furnace, pottery 2nd and 3rd c.
SxNQ XVII (1969), 101–03.

HÉBERVILLE Lug., *Caletes* (Seine-Maritime, F) 49°47′N, 0°47′E
Gallo-Roman foundations; villa.
Cochet, *S.I.,* 435.

HEDGERLEY GREEN Brit., *Catuvellauni* (Buckinghamshire, GB) 51°35′N, 0°36′W
Pottery kiln, mid-2nd c.
Records of Buckinghamshire XIII (1934–40), 252–80.

HELENA Bel., *Atrebates* (?Pas-de-Calais, F)
Sid. Apoll., *Pan. Majorian.,* V.
Minor settlement, site uncertain. Place of the victory of Aëtius over the Franks on the eastern confines of the *civitas* of the Atrebates. ?Allains, Bapaume, Helesmes, Houdain, Lens, Noyelles-Godault, Vis-en-Artois, Vitry-en-Artois?
Dérolez, *Cité des Atrébates*: *RÉA* XLVI (1944), 21: *RN* XLVIII (1966), 517: Delmaire, 1976, 125–6.

HEMSWORTH Brit., *Durotriges* (Dorset, GB) 50°51′N, 2°03′W
Villa, courtyard. Occupied during first three quarters of the 4th c. Baths, mosaics, coins Constantine I to Gratian.
PDNHAFC XXX (1909), 1–12.

HÉNIN-LIÉTARD Bel., *Atrebates* (Pas-de-Calais, F) 50°25′N, 2°56′E
Gallo-Roman foundations, Roman road. *Cf* Noyelles-Godault.
Leduque, *Atrébatie,* 80.

HÉNOUVILLE Lug., *Veliocasses* (Seine-Maritime, F) 49°29′N, 0°58′E
Gallo-Roman foundations; cremated burials.
BSNÉP XXXV, fasc. 4, 99, 149.

HERCVLIS PROMONTORIVM *cf* Hartland Point

HÉRICOURT-EN-CAUX Lug., *Caletes* (Seine-Maritime, F) 49°42′N, 0°42′E
Gallo-Roman foundations at the hamlet of Gréaume; ?temple. On an island in the Durdent, a villa.
Cochet, *S.I.,* 445: *BCASM* I, 287: *Rev de Norm,* 1868; *Album Comm. Ant SI,* IV, 22.

HERIVS FLVMEN *cf* ?river Vilaine

HERMANVILLE-SUR-MER Lug., *Viducasses* (Calvados, F) 49°17′N, 0°19′W
'les rues de Roncheville', possible fort. Coins including a consecration issue of Claudius II, a Constantine and a Byzantine coin of Heraclius.
BSAN XXIX (1913), 226; XLII (1935), 354–55.

HERMES Bel., *Bellovaci* (Oise, F) 49°22′N, 2°15′E
?RATOMAGVS (*qv*).
Minor settlement, 'Rouen', square in plan. Roman roads. Findspot of *CIL* XIII, 3475, with possible reference to *Vicus Ratumagensium.*
Gallia VII (1941), 122: Roblin, 1978, 240–43.

HÉROUVILLE-ST-CLAIR Lug., *Viducasses* (Calvados, F) 49°12′N, 0°19′W
Vicus or minor settlement; various foundations, villa. Flavian and Antonine pottery of la Graufesenque, la Madeleine, Lezoux. 1st and 2nd c.
de Caumont, 1846, 66: *Ann 5 Norm* IV (1967), 317–36: *BSAN* LVIII (1968), 428–30: *Gallia* XXVI (1968), 349–54.

***HESCAMPS-ST-CLAIR** Bel., *Ambiani* (Somme, F) 49°44′N, 1°53′E
Gallo-Roman foundations, tiles, coins.
Leduque, *Ambianie,* 163, n 699.

HESDIN-LE-VIEUX Bel., *Atrebates* (Pas-de-Calais, F) 50°22′N, 2°02′E
?HELENA (*qv*).
Leduque, *Morinie,* 85: *RN* XXXII (1950), 94:

Delmaire, 1976, 340.

HEUDREVILLE-SUR-EURE Lug., *Aulerci Eburovices* (Eure, F) 49°09′N, 1°11′E
Temple in the château park; 'La Londe', villa.
Coutil, II (*Louviers*), 253: Grenier, *Manuel* II.2 (1934), 870.

HIGHGATE WOOD Brit., ?*Catuvellauni/ Londinienses* (Greater London, GB) 51°35′N, 0°09′W
Pottery, at least 8 kilns; two tonnes of pottery c AD 60–120 have been recovered.
LA I.2 (1969), 38–44 through to 300–03.

HIGH HAM Brit., *Durotriges* (Somerset, GB) 51°03′N, 2°49′W
Villa; mosaics, coins of Constantius I.
VCH Somerset I (1906), 328: Rainey, 1973, 94.

HIGH WYCOMBE Brit., *Catuvellauni* (Buckinghamshire, GB) 51°37′N, 0°44′W
Villa, double-corridor. Detached bath-house and other buildings; mosaics. Built late 2nd c, still occupied late 4th c.
Records of Buckinghamshire XVI (1959), 227–57: Rainey, 1973, 94.

HILLION Lug., *Coriosolites* (Côtes du Nord, F) 48°31′N, 2°41′W
CIL XIII, nos 10,025 to 46a.
At 'la Granville', important villa; debris, bas-reliefs in marble. Coins from Nero to Tetricus. At 'l'Hôtellerie' substantial Gallo-Roman foundations; a fibula; terra sigillata REGVLIANVS. At 'Carquitte' three to four kg. of coins of the late Empire; With reference to *CIL* a lead coffin and a glass 'barillet frontinien'.
Habasque, 1834, 433; 1836, 90: *ACN*, 1838, 95–99: *BSCN*, 1868, 125; 1878, 141–49; 1895, 204, 207: *MSCN* I (1883–84), 168: Mechain, P., unpublished.

HINTON ST MARY Brit., *Durotriges* (Dorset, GB) 50°57′N, 2°18′W
Villa, courtyard. Remarkable mosaic with head of Christ, and other mosaics. Occupied c AD 270–400. Mosaic in British Museum.
JRS LIV (1964), 7–14: RCHM *Dorset* III.1 (1970), 117: Rainey, 1973, 95.

HIRFYNYDD Brit., *Silures* (West Glamorgan, GB) 51°43′N, 3°43′W
Signal station.
Nash-Williams, 1969, 140–41.

HOBARROW BAY Brit., *Durotriges* (Dorset, GB) 50°37′N, 2°09′W
Salt working; from late Iron Age into early Romano-British period.
RCHM *Dorset* II part 3 (1970), 613.

HOD HILL Brit., *Durotriges* (Dorset, GB) 50°54′N, 2°12′W
?DVNVM.
Ptol. II,3,13 (Δούνιον).
PNRB 344.
Fort. Situated in a hill fort of the *Durotriges* after this had been attacked and destroyed by *Legio II Augusta* under Vespasian (c 44). Brief occupation. Timber buildings proper to such a fort. Many small finds in British Museum.
Richmond, I. A., *Hod Hill II*, London, 1968.

HOLBOROUGH KNOB Brit., *Cantiaci* (Kent, GB) 51°19′N, 0°24′E
Tumulus; diameter at base c 28 m, height 5.5 m, surrounded by a deep ditch. Cremation burial in a wooden coffin; secondary inhumation in a decorated sarcophagus. Three funerary pits, one containing a folding chair. Nails, pottery, glass, 3rd c.
Arch Cant LXVIII (1954), 1–61.

HOLCOMBE Brit., *Durotriges* (Devon, GB) 50°44′N, 2°58′W
Villa; mosaic and tessellated pavements including octagonal plunge bath. Timber building c AD 80; stone 2nd c; bath-house added 4th c.
Fox, A., *South-West England*, Newton Abbot, 1973, 172: Rainey, 1973, 95: *PDAS* XXXII (1974), 59–161.

HONEYDITCHES *cf* **SEATON**

HOO Brit., *Cantiaci* (Kent, GB) 51°24′N, 0°34′E
Salt working; briquetage, pottery.
Arch Cant LXXX (1965), 273.

HÔPITAL CAMFRONT (L') Lug., *Osismii* (Finistère, F) 48°20′N, 4°14′W
'Keroulé', Gallo-Roman foundations.
BSAF, 1875, 129: du Chatelier, 1907, 113.

HORNENSIS LOCVS *cf* ?Saint-Valéry-sur-Somme

*****HORNOY** Bel., *Ambiani* (Somme, F) 49°51′N, 1°54′E
Temple, 'Le Val d'Aumont', 'La Belle Épine'. 'La bastille de César', camp, Roman road, pits, tomb. 'Le champ de trésor', coin hoard of 235.
Agache, fig. 573: *BSAP*, 1972, 322: Leduque, *Ambianie*, 12: Agache-Bréart, 1975, 78.

HOUDAIN Bel., *Atrebates* (Pas-de-Calais, F) 50°27′N, 2°32′E
?Castellum.
Dérolez, 521.

HOULBEC-COCHEREL Lug., *Aulerci Eburovices* (Eure, F) 49°04′N, 1°22′E
Villa, with cemetery.
Le Prévost, 1862, II, 263: *Congrès archéologique de France* LVI (Évreux, 1889), 87.

HOUPPEVILLE Lug., *Veliocasses* (Seine-Maritime, F) 49°31′N, 1°06′E
Villa, corridor, in the green wood at a place called 'La Butte'.
BCASM X, 195; XX, 33: *BSNÉP* XXXIV, fasc. 2.

HOUVILLE-LA-BRANCHE Lug., *Carnutes* (Eure-et-Loir, F) 48°27′N, 1°38′E
Gallo-Roman foundations, coin hoard.
Nouel, 20, 36.

HUCCLECOTE Brit., *Glevenses* (Gloucestershire, GB) 51°51′N, 2°11′W
Villa, corridor. Baths, mosaics, tiles stamped

R P G; coins chiefly of 4th c; wall plaster with depiction of the end elevation of a house. Built *c* AD 150, occupied into 5th c.
TBGAS LV (1933), 323–76; LXXIX (1960), 159–73; LXXX (1961), 42–9: Rainey, 1973, 97: McWhirr, 1981, 99–100.

HUNTSHAM Brit., *Dobunni* (Herefordshire, GB) 51°51′N, 2°38′W
Villa; in a loop of the river Wye. House, occupied *c* AD 200 to late 4th c; aisled barn used late-3rd to mid-4th c.
TWNFC XXXVII (1961), 179–91: *JRS* LV (1965), 208; LVI (1966), 206.

HURSTPIERPOINT Brit., *Regni* (West Sussex, GB) 50°55′N, 0°11′W
Villa; mosaic, hypocausts. Tile kiln to east.
SxAC XIV (1862), 176–81: *VCH Sussex* III (1935), 58.

IA(N)TINVM then **MELDI** *cf* Meaux

ICKLESHAM Brit., *Regni* (East Sussex, GB) 50°54′N, 0°40′E
Iron workings; 6 furnaces, coins of Hadrian.
SxNQ VI (1937), 247–48.

ICTIS INSVLA
Diod. Sic. V,22,2 (῎Ικτιν).
Plin. *NH* IV, 104 (*Mictim*).
PNRB 487–8.
Possibly to be identified with St Michael's Mount, Cornwall (*cf* Maxwell, I. S., The Location of Ictis, *JRIC* VI, 1972, 293–319; *REA* LXXIII (1971), 403); but there is perhaps a stronger case for supposing that ICTIS should be identified with VECTIS (the Isle of Wight, *qv*).

IFFENDIC Lug., *Redones* (Ille-et-Vilaine, F) 48°08′N, 2°02′W
'les Moulinets', in a field called 'le champ du Clos du Puits' beside the Monet, Gallo-Roman foundations, a square and a round in brick, a hearth.
BSAIV X (1876), 3, 4.

IFFS (Les) Lug., *Redones* (Ille-et-Vilaine, F) 48°17′N, 1°52′W
Gallo-Roman foundations, 'Bourg', at the presbytery near the church. A mass of cement, objects of bronze.
BSAIV V (1859), procès-verbaux, 52; 1861, 80; 1892, procès-verbaux, xxvi, lii.

IFOLD *cf* Painswick

IFORD Brit., *Belgae* (Avon, GB) 51°19′N, 2°18′W
Villa; baths, mosaics, coins of late-3rd and early-4th c.
VCH Somerset I (1906), 300: Rainey, 1973, 97.

ILCHESTER Brit., *Durotriges* (Somerset, GB) 51°00′N, 2°41′W
?LINDINIS.
Rav. Cos. 106.11.
RIB 1672, 1673 (references to *Durotriges Lendinienses*).

PNRB 392.
Town at the crossing of the river Yeo on the road between Bath and Axminster. Occupied from *c* AD 60 to early 5th c; from the 3rd c it was apparently the capital of the northern part of the *Durotriges*, *qv*. A 2nd-c earthen rampart was succeeded by a 4th-c wall enclosing 13 ha. Houses, cemeteries, pottery kilns.
Arch J CVII (1950), 94–5: Rainey, 1973, 97–8: Wacher, 1975, 407–8.

ILCHESTER MEAD Brit., *Durotriges* (Somerset, GB) 51°00′N, 2°42′W
Villa; occupied mid 2nd to late 4th c. Mosaics, pottery, coins 3rd and 4th c.
PSANHS XCVI (1951), 51: *SDNQ* XXVII (1955–60), 80–81: Rainey, 1973, 98.

ÎLE D'AURIGNY *cf* Alderney

ÎLE DE GUERNESEY *cf* Guernsey

ÎLE DE JERSEY *cf* Jersey

ÎLE DE SEIN Lug., *Osismii* (Finistère, F) 48°00′N, 4°53′W
SENA INSVLA.
Mela III, 48.
Plin. *NH* IV, 103 (*Samnis*).
It. M. 509.3 ((*S*)*ina*).
'le Roujou', Gallo-Roman foundations. Terra sigillata, amphorae, coin of Antoninus Pius.
du Chatellier, 1907, 235.

ÎLE D'OUESSANT (USHANT) Lug., *Osismii* (Finistère, F) 48°27′N, 5°04′W
VXISAMA or VXANTIS INSVLA.
Str. I,4,5 (C.64) (Οὐξισάμη).
Plin. *NH* IV, 103 (*Axanthos*).
It. M. 509.3 (*Uxantis*).

ILLOGAN (Magor Farm) Brit., *Dumnonii* (Cornwall, GB) 50°13′N, 5°18′W
Small villa, corridor; tessellated pavement, wall plaster; occupied late 2nd to 3rd c. The only villa known west of Exeter, possibly associated with tin-mining.
Ant V (1931), 494–95: *Ant J* XII (1932), 71–72: *JBAA* XXXIX (1933), 117: Branigan-Fowler, 1976, 171.

INCARVILLE Lug., *Aulerci Eburovices* (Eure, F) 49°14′N, 1°11′E
Villa, 'Le Testelet'.
Coutil, II (*Louviers*), 288: Grenier, *Manuel* II.2 (1934), 871.

INCHEVILLE Lug., *Caletes* (Seine-Maritime, F) 50°01′N, 1°30′E
Cemetery, 4th–5th c; Gallo-Roman cemetery in the valley of the Bresle; small finds. Tile kiln 'Quêne à Leu'. 'Le Camp de Mortagne', oppidum of the *Caletes,* near by.
Cochet, *S.I.,* 331: *BCASM* II, 203; III, 347: *DAG*: *BSPF* XVI (1919), list lxxvii, 196: Agache, fig. 369: Leduque, *Ambianie,* 168, n 785, 786.

INGENA *cf* Avranches

IPING Brit., *Regni* (West Sussex, GB) 51°02′N, 0°48′W

Minor defended settlement, posting station on the road between Silchester and Chichester.
SxAC XCI (1953), 1–3: Cunliffe, 1973, 72: Rodwell-Rowley, 1975, 47, 112.

IRISH SEA
OCEANVS HIVERNICVS.
Ptol. II,2,7; II,3,2 ('Ὠκεανὸς 'Ιουερνικός).
Marcian. II, 42; 44.
PNRB 44.

ISARA FLVMEN *cf* river Oise

ISCA *cf* Caerleon

ISCA DVMNONIORVM *cf* Exeter

ISCA FLVMEN *cf* rivers Exe and Usk

ISCALIS *cf* ?Charterhouse

ISLE OF WIGHT Brit., ?*Belgae* (Isle of Wight, GB) 50°N, 1°W
VECTIS INSVLA.
Plin. *NH* IV, 103 (*Vectis*).
Suet. VIII,4,1 (*insulam Vectem*).
Ptol. II,3,14; VIII,3,11 (Οὐηκτίς).
It. M. 509.2 (*Vecta*).
Paneg. Const. Caes. 15,1 (*Vectam insulam*).
Eutropius, *Brev.* VII,19,1 (Βέκτη).
Rav. Cos. 105.29 (*Vectis*).
Beda *Chron.* 297: *Ecc. Hist. praefatio*; I,3; I,15; IV,13; IV,16; V,19; V,23 (*Vecta*).
PNRB 487–9.
Cf ICTIS INSVLA.

ISLIP Brit., *Catuvellauni* (Oxfordshire, GB) 51°49'N, 1°14'W
Villa, winged corridor. Within rectangular enclosure. Known only from air photographs.
Oxon XXVIII (1963), 89; *Britannia* V (1974), 257.

ISQUES Bel., *Morini* (Pas-de-Calais, F) 50°40'N, 1°39'E
?VICVS DOLVCENSIS (*cf* Halinghen).
?PORTVS AEPATIACI (*qv*).
Minor settlement.
Desjardins, *Géo. Gaule*, I, 370, 400: *CRAI*, 1944, 372: *RÉA* XLVI (1944), 299: Leduque, *Boulonnais*, 56.

ISSY-LES-MOULINEAUX Lug., *Parisii* (Hauts-de-Seine, F) 48°49'N, 2°16'E
Late cemetery.
Duval, 1961, 231, 293: *Gallia* XXVIII (1970), 239: Roblin, 1971, 47, 129, 156, 262.

ITCHEN ABBAS Brit., *Belgae* (Hampshire, GB) 51°06'N, 1°15'W
Villa; mosaics, plaster, bricks, coin of Constantine I.
VCH Hampshire I (1900), 307: Rainey, 1973, 98.

ITCHINGFIELD Brit., *Regni* (West Sussex, GB) 51°03'N, 0°22'W
Tile kiln, ?2nd c. Buildings, flue-tiles, *tegulae, imbrices*. Some tiles from I. used at Alfoldean, *qv*.
SxAC CVIII (1970), 23–38.

IVLIOBONA *cf* Lillebonne

IWADE Brit., *Cantiaci* (Kent, GB) 51°23'N, 0°42'E
Salt workings; evaporation vessels, pottery.
Arch Cant LXXX (1965), 260–65, 273.

IZEL-LES-HAMEAUX Bel., *Atrebates* (Pas-de-Calais, F) 50°19'N, 2°32'E
Gallo-Roman foundations.
Jelski, 143.

JERSEY ?Lug., ?*Unelli* Channel Islands (GB) ÎLE DE JERSEY Les Îles Anglo-Normandes 49°13'N, 2°08'W
?ANDIVM INSVLA.
It. M. 509.3.
PNRB 181.
T. D. Kendrick, *Archaeology of the Channel Islands I, The Bailiwick of Guernsey*, 1928, 16: J. J. Hawkes, *Archaeology of the Channel Islands II, The Bailiwick of Jersey*, n.d., pub. 1937, 18. Gallo-Roman temple at the foot of Pinnacle rock, pottery 2–4 c; *Annual Bulletin of the Société Jersiaise* 15, 1959–60, 226–34: Johnston, 1977, 31. At Quennevais, coin hoard of *c* 400 coins, 3–4 c; J. J. Hawkes, *op cit*, 129–31. At Île Agois, coin hoard of 17 coins Valerian to Postumus, 253–268; *Annual Bulletin of the Société Jersiaise*, 1977.

JORDON HILL Brit., *Durotriges* (Dorset, GB) 50°38'N, 2°26'W
Romano-Celtic temple, active late 4th and early 5th c. Many small finds including coins; remarkable ritual shaft.
Coles, J. and Simpson, D. D. A., eds, *Studies in Ancient Europe*, Leicester, 1968, 266: RCHM *Dorset* II part 3 (1970), 616.

JORT Lug., *Viducasses* (Calvados, F) 48°58'N, 0°04'W
Vicus or minor settlement; foundations, pits, numerous small finds; coarse pottery and terra sigillata including CRISPINI M. VAPVSO. OF PATRICI. OF SEVER. PRIMVS F. GEMINI. GERMANI. BVTRIO. SAMILLI M.
de Caumont, 1831, 231; 1850, 365: *MSAN* VI (1831–33), 440–44; XX (1855), 325–44: *BSAN* LII (1955), 111: *Gallia* XXVIII (1970), 272.

JOUARS-PONTCHARTRAIN Lug., *Carnutes* (Yvelines, F) 48°47'N, 1°54'E
DIVODVRVM.
It. A. 384.6 (*Dioduro*).
Minor settlement.
Duval, 1961, 118, 244, 245.

JOUÉ DU PLAIN Lug., *Esuvii* or *Sagii* (Orne, F) 48°41'N, 0°07'W
'La Course', *vicus* or minor settlement; aqueduct. Tegulae, sword pommel.
MSAN IX (1835), 461–62: *BSHAO* III (1884), 150.

JUAYE-MONDAYE Lug., *Baiocasses* (Calvados, F) 49°13'N, 0°41'W
Aqueduct leading towards Bayeux; wells. Two coins of Constantine, three of Constantius II, one

Claudius II, one Valens; mid 4th c.
de Caumont, 1857, 372, 384: *BSAN* II (1862), 118–21.

JUBLAINS Lug., *Aulerci Diablintes* (Mayenne, F) 48°17′N, 0°30′W
NOVIODVNVM DIABLINTVM.
Ptol. II,8,7 (Νοιόδουνον).
?*Tab. Peut.* (*Nudionnum*). *Cf* Sées. *Cf* AVLERCI.
Town, capital of the CIVITAS DIABLINTVM. Theatre, temple, baths, late Roman *burgus*, early Christian church (succeeding baths).
Grenier, *Manuel* I (1931), 454–63; III (1958), 964–6; IV (1960), 777–86: *Gallia* XII (1954), 170–2; XIII (1955), 164–5; XVII (1959), 354–5: *Ann Bret* LXV (1958), 73–85: Lantier, R., *BACTH* 1943–55, 177.

JUILLÉ Lug., *Aulerci Cenomani* (Sarthe, F) 48°15′N, 0°07′E
Gallo-Roman foundations. On the left bank of the Sarthe numerous important substructures, an octagonal building of brick, internal diameter 4.55 m – temple, or ?tower.
Bouton: Ledru, 1911, 106–9, 354.

*****JUMEL** Bel., *Ambiani* (Somme, F) 49°45′N, 2°21′E
Gallo-Roman foundations; sarcophagus, tiles, pottery, coins.
Agache, fig. 163, 326, 480, 639: Leduque, *Ambianie,* 149 n 445: Agache-Bréart, 1975, 79: Agache, 1978, 316, 429.

KELVEDON Brit., *Trinovantes* (Essex, GB) 51°50′N, 0°42′E
CANONIVM.
It. A. 480.5 (*Canonio*).
Tab. Peut. (*Caunonio*).
PNRB 296–7.
Major settlement, succeeding a fort, on the road between London and Colchester. Potters' workshop, cemetery, Romano-Celtic temple, traces of dwellings.
VCH Essex III (1963), 149: *JRS* LIX (1969), 223, 245: *Britannia* I (1970), 52, 70; V (1974), 442–3: Dunnett, 1975, *passim*: Rodwell-Rowley, 1975, 85–101.

KERLAZ Lug., *Osismii* (Finistère, F) 48°05′N, 4°16′W
'Lanévry', salt manufactory on the Bay of Douarnenez near the road between Douarnenez and the Crozon peninsula. Coins of Marcus Aurelius, Tetricus, Constantine. Terra sigillata of 2nd and 3rd c, Argonne of 4th c.
Gallia XXX (1972), 213.

KERMOROC'H Lug., *Osismii* (Côtes du Nord, F) 48°37′N, 3°12′W
South of the town numerous Gallo-Roman foundations and much building debris.
MSCN I (1883–84), 19–20.

KERNILIS Lug., *Osismii* (Finistère, F) 48°34′N, 4°25′W

CIL XIII 9016.
Milestone of Claudius I on the road between Kérilien and l'Aberwrac'h.
MSHAB, 1956, 6–9.

KERNILIS Lug., *Osismii* (Finistère, F) 48°34′N, 4°25′W
'Kerbrat Huella', Gallo-Roman foundations which might be those of a villa, near the road between Kérilien and Lannilis.
BSAF, 1875, 130.

KESTON Brit., *Cantiaci* (Kent, GB) 51°21′N, 0°01′E
Villa, corridor; baths, mausoleum, funerary pits, tombs.
Arch Cant LXXXII (1967), 184–91: *JRS* LIX (1969), 232: *KAR* XI (1968), 10–14; XXI (1970), 21–23.

KEYNSHAM Brit., *Belgae* (Avon, GB) 51°25′N, 2°31′W
Villa; large courtyard. Many mosaics, coins *c* AD 265 to 375.
Arch LXXV (1926), 109–138: Rainey, 1973, 99–101.

KIMMERIDGE BAY Brit., *Durotriges* (Dorset, GB) 50°37′N, 2°08′W
Shale working.
RCHM *Dorset* II part 3 (1970), 601.

KINGSCOTE Brit., *Regni* (West Sussex, GB) 51°06′N, 0°03′W
Iron workings; slag heaps; pottery of late 1st to late 3rd c.
VCH Sussex III (1935), 30.

KING'S WESTON Brit., *Belgae* (Avon, GB) 51°29′N, 2°40′W
Villa; occupied late 3rd to late 4th c. Baths, mosaics.
TBGAS LXIX (1950), 5–58: Rainey, 1973, 101–02.

KINGS WORTHY Brit., *Belgae* (Hampshire, GB) 51°06′N, 1°18′W
Villa; 4th-c occupation. Mosaics, hypocausts.
JRS XV (1925), 243.

KIT'S COTY Brit., *Cantiaci* (Kent, GB) 51°19′N, 0°30′E
Temple on the site of an Iron Age temple; abandoned early 5th c.
Lewis, 1966, 124: *VCH Kent* III (1932), 104.

LADY DOWN *cf* Chilmark
*****LAHOUSSOYE** Bel., *Ambiani* (Somme, F) 49°57′N, 2°29′E
Gallo-Roman foundations (?villa), tiles, pottery, coins, bas-relief, 'Le Templier'.
Agache, fig. 226, 239: *MSAP* III (1858), 151: Leduque, *Ambianie,* 135: Vasselle, 320: Agache-Bréart, 1975, 81. Agache, 1978, 246, 253, 256, 257, 269, 270, 332, 438.

LAKE Brit., *Durotriges* (Dorset, GB) 50°48′N, 2°00′W

A vexillation fortress of 11.7 ha with two phases of occupation, both of mid-1st c., lying on road from Poole Harbour to Badbury.
PDNHAS XCI (1969), 188: *C.Arch* II (1969–70), 188: *Britannia* I (1970), 299; II (1971), 281; III (1972), 346; V (1974), 7, 455; XI (1980), 391; XIII (1982), 384.

LAMBOURNE'S HILL Brit., *Belgae* (Hampshire, GB) 51°15′N, 1°35′W
Villa, corridor. Hypocausts, coins of late 3rd and early 4th c.
VCH Hampshire I (1900), 294.

LAMYATT BEACON Brit., *Durotriges* (Somerset, GB) 51°07′N, 2°28′W
Romano-Celtic temple; annexes to east and west, small building to north. Numerous bronze statuettes, votive spearheads, late Roman belt-plate, knee of life-size statue, etc. c 2500 coins.
JRS LI (1961), 187: *Britannia* V (1974), 452.

LANCIEUX Lug., *Coriosolites* (Côtes du Nord, F) 48°37′N, 2°09′W
CIL XIII, 8994.
Milestone, 4th c.
BSCN XXIX (1891), 191; XXXV (1897), 234; XLVII (1909), 80.

LANCING Brit., *Regni* (West Sussex, GB) 50°51′N, 0°20′W
Romano-Celtic temple, situated over an Iron Age shrine which stood within an oval fenced temenos. Bronze Age and Iron Age pottery and coins suggest a long history which extended into the Romano-British period. Many small finds, coins from 1st and 2nd c.
VCH Sussex III (1935), 59: *SxAC* LXXXI (1940), 158–69; CXIX (1981), 37–55; Lewis, 1966, *passim*: Cunliffe, 1973, 108–9.

LANDERNEAU Lug., *Osismii* (Finistère, F) 48°27′N, 4°15′W
Vicus, meeting point of several roads, ford. Numerous foundations. Bronze ring. Sesterces of Hadrian and Antoninus Pius, antoninianus of Tetricus.
Bull Soc Brest, 1869: de Courcy, 1906: *BSAF*, 1973, 42–44.

LAND'S END Brit., *Dumnonii* (Cornwall, GB) 50°04′N, 5°43′W
ANTIVESTAEVM sive BELERIVM PROMONTORIVM.
Diod. V.21.3, V.22.1 (ἀκρωτήριον Βελέριον).
Ptol. II.3.2. ('Αντιουέσταιον ἄκρον τὸ καὶ Βολέριον).
PNRB 252, 266.

LANFAINS Lug., *Coriosolites* (Côtes du Nord, F) 48°21′N, 2°55′W
'Carestiemble', important villa.
MSCN I (1883–84), 249.

LANGENHOE Brit., *Trinovantes* (Essex, GB) 51°48′N, 0°59′E
Salt workings, briquetage, Iron-Age and R-B pottery.
Nenquin, 1961, 87: *VCH Essex* III (1963), 32–4, 151.

LANGONNET Lug., ?*Veneti* (Morbihan, F) 48°06′N, 3°30′W
At 'la Trinité', under the presbytery, a many-roomed villa; building debris including white marble. At 'Stang Yan' a villa 21 m by 9 m in extent; building debris, terra sigillata, coins of Trajan. At 'Kergaradec' 'Minex-Bloch', 'Kervruc', Gallo-Roman foundations, tegulae, great numbers of bricks. At 'Guernegal-Castel', tile- and brick-kilns, numerous waste-heaps.
ASPM, 1865, 572: Fouquet, 1868, 78: Le Mené, 1891, 402: *BSAF*, 1895, 115 sqq: *BSAL* I (1906), 205: *BSPM*, 1910, 47–48: Leroux, A., 1–47.

LANGUEUX Lug., *Coriosolites* (Côtes du Nord, F) 48°30′N, 2°43′W
'les Grèves', Gallo-Roman foundations, marble pavements, coin of Gallienus.
MSCN I (1883–84), 169.

LANISCAT Lug., *Osismii* (Côtes du Nord, F) 48°14′N, 3°07′W
Gallo-Roman foundations in 'le Blavet'; numerous coins.
MSCN I (1883–84), 504.

LANVALAY Lug., *Coriosolites* (Côtes du Nord, F) 48°27′N, 2°02′W
'la Muraille de l'Oeuvre', foundations of a Roman bridge over the Rance between Corseul and Dol.
BSCN XLVII (1909), 18.

LARCHANT Lug., *Senones* (Seine-et-Marne, F) 48°17′N, 2°36′E
Gallo-Roman foundations, sanctuary, cemetery, coin hoard.
Grenier, *Manuel* IV.2 (1960), 715: Nouel, 20, 27, 36.

LATIMER Brit., *Catuvellauni* (Buckinghamshire, GB) 51°41′N, 0°33′W
Villa. Timber building of c AD 80 lasting to c 120; replaced by stone house which lasted through several reconstructions until c AD 400, then occupied through 5th c. Many finds of all periods; burials of a pig, a sheep and a calf in pits c 450.
Branigan, K., *Latimer*, Chesham, 1971.

***LAUCOURT** Bel., *Ambiani* (Somme, F) 49°40′N, 2°46′E
Gallo-Roman foundations, 'Fort de Chessoy, le Vieux Catiau'.
Vasselle, 322: Agache-Bréart, 1975, 82: Agache, 1978, 438.

LÉALVILLERS Bel., *Ambiani* (Somme, F) 50°04′N, 2°31′E
Cemetery, pottery, coins.
Leduque, *Ambianie*, 30.

LÉBISEY Lug., *Viducasses* (Calvados, F) 49°13′N, 0°20′W
Villa, corridor, tripartite plan. Columns of Caen stone.
MSAN X (1837), 318–25: de Caumont, 1838, 139–41: Grenier, *Manuel* II.2 (1934), 800.

LECOUSSÉ Lug., *Redones* (Ille-et-Vilaine, F) 48°22′N, 1°13′W

'Marvaize', villa; foundations, bricks, fragments of terra sigillata; statuette of Venus Anadyomene.
BSAIV XXXIX (1909–10); LIV (1928), 52.

LEGEDIA *cf* Avranches

LEHON Lug., *Coriosolites* (Côtes du Nord, F) 48°26′N, 2°03′W
'Château-Fort', Gallo-Roman foundations.
BSCN XLVII (1909), 19.

LEMANIS PORTVS *cf* Lympne

LENAULT Lug., *Viducasses* (Calvados, F) 48°56′N, 0°37′W
Gallo-Roman foundations, ?villa. Two cisterns 2.50 m deep. Bricks, tegulae, charcoal; coarse pottery and terra sigillata PATERNVS. CORB. 2nd c.
BSAN VII (1875), 378.

LENS Bel., *Atrebates* (Pas-de-Calais, F) 50°26′N, 2°50′E
Cemetery.
Leduque, *Atrébatie*, 73.

LETHUIN Lug., *Carnutes* (Eure-et-Loir, F) 48°22′N, 1°52′E
Minor settlement, ?villa, on the road between Beauvais and Blois. Foundations, coins.
Jalmain, 95: Nouel, 20.

LEVCA FLVMEN *cf* river Loughor

LEVCARVM *cf* Loughor

LEVCOMAGVS *cf* ?East Anton

LEULINGHEM Bel., *Morini* (Pas-de-Calais, F) 50°44′N, 2°10′E
Gallo-Roman cemetery, 'Uzelot', 'Le chemin des Morts'.
Leduque, *Boulonnais*, 86.

LEXDEN, THE MOUNT Brit., *Trinovantes* (Essex, GB) 51°53′N, 0°51′E
Tumulus, late Iron Age; diameter at base 31m, height 4.5 m; pottery, tiles.
TEAS XII (1912), 186–92: Hull, 1958, 252.

LEXOVII
Caes. *BG* III, 9, 11, 17, 29; VII, 75.
Str. IV,1,14 (C.189) (Ληξοβίους); IV,3,5 (C.194) (Ληξοούιοι).
Plin. *NH* IV, 107.
Ptol. II,8,2 and 5 (Ληξούβιοι).
Not. Gall. II, 7 (*civitas Lexoviorum*).
Oros. VI, 8.
Greg. Tur. *HF* VI, 36 (*Lixoensis episcopus*).
CIL XIII.1, p 502.
Tribe of Celtic Gaul, included in Lugdunensis (later Lugdunensis II). Oppidum: Cambremer. Capital: NOVIOMAGVS LEXOVIORVM, later known as LEXOVIIS (Lisieux, *qv*).
BSNÉP XXVII (1927–9), 139–166.

LÉZARDRIEUX Lug., *Osismii* (Côtes du Nord, F) 48°46′N, 3°06′W
'Castel-ar-Hoc', Gallo-Roman foundations; debris, red potsherds with reliefs, coins.
MSCN I (1883–84), 291.

LICKFOLD *cf* **WIGGONHOLT**

*****LICOURT** Bel., *Ambiani* (Somme, F) 49°48′N, 2°53′E
Temple, 'Le Catelet, La Natetière'.
Agache, 1972, 322: Agache-Bréart, 1975, 82: Agache, 1978, 431.

LIERCOURT-ÉRONDELLE Bel., *Ambiani* (Somme, F) 50°02′N, 1°54′E
Temple; bases, shafts, capitals, tiles, pottery, coins. Oppidum of the *Ambiani* ('Les Castelis'), 'Camp de César' with attached Roman camp.
Agache, fig. 29, 48, 49, 64, 78–82, 113, 157, 161, 183, 376–80: *RN* CLXXVI (1962), 324–36: *BSPF* XVII (1920), list LXXIX, 56: *Gallia* XXI (1963), 342; XXIII (1965), 297: Leduque, *Ambianie*, 79, n 819: Vasselle, 322: Agache-Bréart, 1975, 82–83.

LIÉVIN Bel., *Atrebates* (Pas-de-Calais, F) 50°25′N, 2°46′E
Cemetery, 4th–8th c. ?*Vicus*.
Dérolez, 512, 521, 531: *RN* XLIX (1967), 741.

LILAS (Les) Lug., *Parisii* (Seine-St-Denis, F) 48°53′N, 2°26′E
Gallo-Roman foundations; cellar or vault with statuette of Minerva, amphora.
Duval, 1961, 232: Roblin, 1971, 230.

LILLEBONNE Lug., *Caletes* (Seine-Maritime, F) 49°31′N, 0°33′E
IVLIOBONA.
Ptol. II,8,5 ('Ιουλιόβονα).
It. A. 382.1, 384.13, 385.1.
Tab. Peut.
CIL XIII.1.1, p 513, nos 3220–3245.
Town and port, capital under the early Empire of the CIVITAS CALETVM. Forum, amphitheatre (early 1st c) adapted to theatre (mid 2nd c, enlarged late 2nd or early 3rd c), baths, aqueduct, houses, mosaics (end of 3rd c), quays. Late Roman castellum near the former forum; bastion. Important cemetery at 'la Côte du Catillon'. Altars, sculpture, small finds in the museum at Rouen.
Rever, F., *Mémoire sur les ruines de L.*, Évreux, 1821: Cochet, *S.I.*, 396–416; *Rép.*, 131, 135: Blanchet, *Enceintes*, 35, 279: *RA*, 1913, 184–208: Vesly, L., Le castrum de Juliobona, *Bull Soc émul SI*, 1914: Deglatigny, 1931, 111–29: Grenier, *Manuel* II.2 (1934), 525; III.2 (1958), 891–8; IV.1 (1960), 345–9: *Gallia* XVII (1959), 336; XX (1962), 427; XXII (1964), 292: *RSSHN* XIV (1959), 36–40; XXVI (1962), 3; XL (1965), 65–71: *BCASM* XXV (1964–5), 254–8: *BSNAF*, 1968 (70), 121–132; 1970 (72), 217.

LIMES (Cité de) *cf* Bracquemont, Neuville-lès-Dieppe.

LIMÉSY Lug., *Caletes* (Seine-Maritime, F) 49°37′N, 0°56′E
Minor settlement, numerous foundations, 'le Champ de Trêsor'; hoard of coins of the *Caletes*.
Cochet, *S.I.*, 179; *BCASM* IX (1891), 78.

LIMNERSLEASE Brit., ?*Regni* or *Atrebates* (Surrey, GB) 51°13′N, 0°38′W
Villa, double corridor. Baths. Pottery of 2nd and 3rd c, three coins of 4th c.
SyAC XXVIII (1915), 41–50.

LIMOGES-FOURCHES Lug., *Parisii* (Seine-et-Marne, F) 48°37'N, 2°40'E
Villa.
Duval, 1961, 233.

LIMAS-MONTLHÉRY Lug., *Parisii* (Essonne, F) 48°38'N, 2°16'E
Major settlement on the road between Paris and Orléans; cemetery.
Duval, 1961, 116, 232, 243.

LINDINIS *cf* ?Ilchester

LINGREVILLE Lug., *Unelli* (Manche, F) 48°57'N, 1°32'W
Gallo-Roman foundations; terra sigillata and coarse pottery in a layer of turf; a fragment of a Venus Anadyomene.
BSAN LV (1959–60), 7.

LINTOMAGVS *cf* Montreuil-sur-Mer (?or Brimeux)

LINWOOD Brit., *Belgae* (Hampshire, GB) 50°53'N, 1°43'W
Pottery; New Forest. Remarkable jar cover with incised design, female deity, radiate god and horseman.
Ant J XVIII (1938), 113–36: *PHFC* XXIII (1965), 29–45: Ross, 1967, 216.

LION-SUR-MER Lug., *Viducasses* (Calvados, F) 49°18'N, 0°19'W
Small villa; tiles, bricks, heaps of shells. Burials in the marsh.
BSNÉP XXIX (1932–33), 61–80.

LISIA INSVLA *cf* ?Guernsey

LISIEUX Lug., *Lexovii* (Calvados, F) 49°08'N, 0°14'E
NOVIOMAGVS LEXOVIORVM, later LEXOVIIS.
Ptol. II,8,2 (Νοιόμαγος).
It. A. 385.3 (*Noviomago*).
CIL XIII, 1593, 2153, 2212, 3177–3182, 8990.
Wuilleumier 342.
Cf LEXOVII.
Town, fort and port, capital of the CIVITAS LEXOVIORVM. Roads, aqueducts, theatre. votive column, many houses; cinerary urns and sarcophagi; pottery, coins and small finds from 1st to 4th c.
MSAN VI (1831–3), 387–9; XVII (1847), 285–94: *BSAN* VII (1875), 288–9; XXXV (1924), 600–09; XXXVIII (1930), 519–20; LIII (1955–56), 169–96; LV (1958), 466–75: Desjardins, *Géo. Gaule* I, 339; *MSNAF* XXXVII (1876), 89–103: Blanchet, *Enceintes,* 39: *BACTH* LXXIII (1911), 322, pl xxiii: Espérandieu III (1911), 3047, 3050: *Études Lexoviennes,* 1928, 257–93: *BSNÉP* XXVII (1930), 139–66: Doranlo, R., *Epigraphie lexovienne,* 1931: *Revue des Établissements Anciennes,* 1932, 159–81: Grenier, *Manuel* III.2 (1958), 917–20: *CRAI,* 1959, 338–46: *Gallia* XVII (1959), 326; XX (1962), 421; XXIV (1966), 257; XXVI (1968), 362; XXVIII (1972), 338: *Forum* II (1972), 4.

LISTERCOMBE BOTTOM Brit., *Dobunni* (Gloucestershire, GB) 51°48'N, 1°54'W
Villa; baths, hypocausts.
JRS XXI (1931), 240.

LITANOBRIGA *cf* ?Saint-Maximin

LITTLECOTE Brit., *Atrebates* (Wiltshire, GB) 51°26'N, 1°34'W
Villa; mosaics.
VCH Wiltshire I (1957), 98: Rainey, 1973, 110. *Britannia* XII (1981), 1–5, 360–1; XIII (1982), 387–9.

LITTLEHAMPTON Brit., *Regni* (West Sussex, GB) 50°49'N, 0°31'W
Villa, corridor. Tiles, *tesserae,* pottery. Occupied 1st to 3rd c.
ANL II (1949–50), 207.

LITTLE LONDON (Hampshire) *cf* Pamber

LITTLE MILTON Brit., *Catuvellauni* (Oxfordshire, GB) 51°42'N, 1°06'W
Villa, corridor. Outbuildings, enclosing ditch.
JRS XL (1950), 102; XLIII (1953), 94.

LITTLETON Brit., *Durotriges* (Somerset, GB) 51°04'N, 2°44'W
Villa, courtyard. Baths, mosaics. Site occupied in 1st c; villa built early 4th c.
VCH Somerset I (1906), 323–4: Rainey, 1973, 110.

LITVS SAXONICVM
Not. Dig. Occ. XXVII–XXVIII.
Late Roman system of coastal defences, commanded by the *Comes Litoris Saxonici. Cf* Boulogne, Bradwell, Carisbrooke Castle, Dover, Lympne, Pevensey, Portchester, Reculver, Richborough, Walton Castle.
White, D. A., *Litus Saxonicum,* University of Wisconsin, Madison, 1961: Will, E., Remarques sur la fin de la domination romaine dans le N. de la Gaule, *RN* XLVIII (1966), 525: Johnston, 1977: Johnson, 1979.

LIZARD (THE) Brit., *Dumnonii* (Cornwall, GB) 49°57'N, 5°13'W
DVMNONIVM SIVE OCRINVM PROMONTORIVM.
Ptol. II,3,2 (Δαμνόνιον τὸ καὶ Ὄκρινον ἄκρον).
Marcian II.45 (τὸ Δάμνιον ἄκρον τὸ καὶ Ὄκριον).
PNRB 344, 429.

LLANDOVERY Brit., *Silures* (Dyfed, GB) 51°59'N, 3°49'W
?ALABVM.
Rav. Cos. 106.26.
PNRB 242.
Fort with four phases of building, first in timber and then in stone, between the Flavian period and *c* AD 160. Walls enclose *c* 2.4 ha. Possible *mansio.*
Nash-Williams, 1969, 95–6.

LLANHAMLACH *cf* Millbrook Farm

LLANTWIT MAJOR Brit., *Silures* (South Glamorgan, GB) 51°25'N, 3°30'W
Villa, courtyard. Mosaics, etc. Founded mid-2nd c, *floruit c* 350, abandoned late 4th c.

Arch Camb CII (1953), 89–163: Rainey, 1973, 111–12: *Britannia* V (1974), 225–50.

LOCKING Brit., *Belgae* (Avon, GB) 51°21'N, 2°55'W

Villa. Bath-building. Pottery 3rd to late 4th c.

PSANHS XCV (1950), 173: *JRS* XLVIII (1958), 146.

LOCKLEYS Brit., *Catuvellauni* (Hertfordshire, GB) 51°50'N, 0°12'W

Villa, winged corridor. Early 1st c a Belgic house on the site; villa built *c* AD 70, enlarged mid 2nd c; burned late 3rd c. Thereafter a small building on the site until the 4th c.

Ant J XVIII (1938), 339–76: Todd, 1978, 33–58, *passim*.

LOCRONAN Lug., *Osismii* (Finistère, F) 48°06'N, 4°12'W

'la Motte', Gallo-Roman foundations which might be those of a temple. Engraved bronze plaque, bronze lamp.

du Fretay, 1898, 103–06.

LOCVS QVARTENSIS SIVE HORNENSIS *cf* ?Saint-Valéry-sur-Somme

LOGES (Les) Lug., *Caletes* (Seine-Maritime, F) 49°42'N, 0°17'E

Gallo-Roman foundations, cemetery 2nd to 3rd c, at 'Bois des Loges'.

Cochet, *S.I.*, 373.

LOGUIVY-PLOUGRAS Lug., *Osismii* (Côtes du Nord, F) 48°31'N, 3°30'W

'Manach-ru', Gallo-Roman foundations; bronze axe, querns, potsherds.

MSCN I (1883–84), 323.

LONDE (La) Lug., *Lexovii* (Seine-Maritime, F) 49°18'N, 0°58'E

Temple, near le Vivier Gamelin: villa at la Butte Colas: group of ruins in the forest near the farm Saint-Nicolas.

Vesly, 15, 20: Deglatigny, 1931, 73–5: Grenier, *Manuel* II.2 (1934), 770.

LONDINIÈRES Lug., *Caletes* (Seine-Maritime, F) 49°50'N, 1°54'E

Gallo-Roman foundations at 'Le Pré-des-Préaux' and 'le Terroir des Fossés': late-Empire cemetery in the modern cemetery.

Cochet, *S.I.*, 529: *BCASM* XII, 31.

LONDINIVM, AVGVSTA *cf* London

LONDON Brit., *Londinienses* (Greater London, GB) 51°31'N, 0°05'W

LONDINIVM, AVGVSTA.

Tac. *Ann.* XIV, 33 (*Londinium*).

Ptol. I,15,7; II,3,12; VIII,3,6 (Λονδίνιον).

It. A. 471–480, *passim* (*Londinio, Lundinio*).

Paneg. Const. Caes. 17,1 (*oppidum Londiniense*).

Conc. Arel. a. 314 (*civitate Londenensi*).

Amm. Marc. XX,1,3; XXVII,8,7; XXVIII,3,1 (*Lundinium, Augusta*).

Not. Dig. Occ. XI, 37 (*Praepositus thesaurorum Augustensium*).

Steph. Byz. (Λινδόνιον).

Rav. Cos. ?106.38 (*Landini*); 106.50 (*Londinium Augusti*).

Beda *Chron.* 532; *Ecc. Hist.* praef.; I,29; II,3,4,6,7; III,7; IV,6,11,12,22; V,23 (*Lundonia*).

CIL VII, 1235 (*Lon*).

EE IX, 1372 =*JRS* XII (1922), 283 (*Londini ad fanum Isidis*).

JRS XLIV (1954), 108; L (1960), 108–11 (*Londinio*).

RIB 1–40.

Medaillon d'Arras (LON).

PNRB 260, 396–8.

Large and important town, centre of the road network of Roman Britain, at the lowest bridge crossing of the river Thames. No direct evidence for pre-Roman settlement, nor for an early Roman fort although this has often been postulated. Town established *c* AD 50, destroyed by Boudica AD 60, gradually rebuilt over the next two decades. Residence of the procurator of Britain probably before AD 60, of the governor (*legatus Augusti*) from late 1st c. Great fire in Hadrian's reign, followed by period of stagnation. Renewed building early 3rd c, when L. became the capital of *Britannia Superior*; in 4th c, renamed *Augusta*, it was probably capital of *Maxima Caesariensis*, and seat of the *Vicarius Britanniarum* and of a bishop. At Leadenhall Market, small forum and basilica built *c* 75, greatly enlarged early 2nd c; at Cannon Street Station, residence of the governor; at Cripplegate, 2nd c fort. Mithraeum, temples of Isis and Jupiter known from inscriptions, monumental arch at Blackfriars, baths, quays, houses, shops, mosaics, abundance of pottery, coins and small finds. Walls built *c* 200, enclosing 132 ha; riverside wall added 4th c. Bridge over the Thames linking L. with an extensive suburb at Southwark. Museum of London; British Museum.

LA, passim: TLMAS, passim: Merrifield, R., *The Roman City of London*, L., 1965; *Roman London*, L., 1969: Grimes, W. F., *The Excavation of Roman and Medieval London*, L., 1968: Rainey, 1973, 112–5: Wacher, 1975, 87–103: *Ant J* LVII (1977), 31–66: Marsden, P., *Roman London*, L., 1980: *Britannia* V (1974), 445–6; VI (1975), 265–71; VII (1976), 345–52; VIII (1977), 408–10; IX (1978), 452–5; X (1979), 311–8; XI (1980), 379–82.

LONG Bel., *Ambiani* (Somme, F) 50°02'N, 1°59'E

Temple; ?Gallo-Roman quay.

Cochet, *S.I.*, 529: Vasselle, 322: Agache-Bréart, 1975, 84: Agache, 1978, 195.

LONGJUMEAU Lug., *Parisii* (Essonne, F) 48°42'N, 2°18'E

NONGEMELLVM (XIth c).

Cemetery 4th to 5th c, 'Champtier de la Porte d'Orléans'.

Duval, 1961, 78: *Gallia* XXI (1963), 355; XXX (1972), 301: Roblin, 1971, 35, 99, 167, 290.

LONGSTOCK Brit., *Belgae* (Hampshire, GB) 51°07'N, 1°31'W

Villa; baths, mosaic, coin of Constantine I.

PHFC IX (1922), 290.

LONGUEAU Bel., *Ambiani* (Somme, F) 49°52′N, 2°20′E
Milestone (AD 305–307), architectural fragments, weapons, statues; grave stones with inscriptions, burials.
Leduque, *Ambianie,* 137, 142.

LONGUEIL-STE-MARIE Bel., *Bellovaci* (Oise, F) 49°21′N, 2°43′E
Gallo-Roman foundations; 'L'Abbaye', mosaics.
Gallia Sup. X.1.1 (1957) no 58.

LORÉ Lug., *Aulerci Diablintes*? (Orne, F) 48°29′N, 0°34′W
'Gué de Loré', 'ruisseau du Bois-Frangées', *vicus* or minor settlement on the road between Jublains and Vieux. Posting station, houses. Coins of the early Empire; pottery kiln, small finds. 1st to 4th c.
PBN II (1959), 148–55: *Ann Norm* IX (1959), 322–24: *Gallia* XX (1962), 425.

LORREZ-LE-BOCAGE Lug., *Senones* (Seine-et-Marne, F) 48°14′N, 2°54′E
Gallo-Roman foundations, aqueduct, bas-relief of Apollo, 'La Cave-aux-Fées'.
Nouel, 19, 26, 29.

LOTVM (LOIVM, LOGIA) *cf* Caudebec-en-Caux

LOUGHOR Brit., *Silures* (West Glamorgan, GB) 51°40′N, 4°04′W
LEVCARVM.
It. A. 484.1 (*Leucaro*).
PNRB 388–9.
Fort and minor settlement; timber buildings, wall plaster, pottery of AD 75–130.
Britannia III (1972), 300; V (1974), 401: *Arch Camb* CXXII (1974), 99–146.

LOUGHOR, river Brit., *Silures* (West Glamorgan, GB)
LEVCA FLVMEN.
Rav. Cos. 108.29.
PNRB 388.

LOUVIERS Lug., *Aulerci Eburovices* (Eure, F) 49°13′N, 1°10′E
Temple, Forêt de L. Fanum des Buis.
Vesly, 25.

LOUVILLE-LA-CHENARD Lug., *Carnutes* (Eure-et-Loir, F) 48°19′N, 1°47′E
Gallo-Roman foundations.
Air photograph, Jalmain.

LOUVRES-MONTMÉLIANT Lug., *Parisii* (Val-d'Oise, F) 49°02′N, 2°30′E
Shrine, ?Mercury.
Duval, 1961, 232, 244: Roblin, 1971, 15, 33, 100, 238.

LOWER SWELL Brit., *Dobunni* (Gloucestershire, GB) 51°56′N, 1°44′W
Villa, plan uncertain. Pottery, tiles, coins.
TBGAS VII (1882–83), 72–75.

LOW HAM Brit., *Durotriges* (Somerset, GB) 51°03′N, 2°48′W
Villa; timber house of 2nd c succeeded by stone-built villa enlarged late 3rd c and again *c* 330; abandoned *c* 367. Baths, hypocausts, mosaics, incl. one of Dido and Aeneas.
PSANHS XCII (1946), 25–28: *SDNQ* XXV (1947–50), 1–6, 61–64, 141–43: Rainey, 1973, 115–17: *Britannia* VII (1976), 358–9.

LUCÉ Lug., *Carnutes* (Eure-et-Loir, F) 48°26′N, 1°28′E
Gallo-Roman foundations, mosaics, sculpture.
Gallia XIII (1955), 161.

LUC-SER-MER Lug., *Viducasses* (Calvados, F) 49°19′N, 0°21′W
Vicus or minor settlement with port, 'la Brèche du Corps de Garde'. Various foundations including those of a house or a potter's workshop; tegulae and imbrices, small finds including coins of Hadrian; la Graufesenque pottery signed SESTIVS. 1st c.
BSAN XXIII (1913), 257; XXXI (1916), 330–7; XXXVIII (1930), 536–8; XLIV (1957), 372, 387.

LUFTON Brit., *Durotriges* (Somerset, GB) 50°57′N, 2°42′W
Villa, corridor. Octagonal bath-house, many mosaics; pottery of 4th c.
PSANHS XCVII (1952), 91–112; CXVI (1971–72), 59–77: Rainey, 1973, 117–18.

LVGVDVNVM *cf* BROCKLEY HILL

LULLINGSTONE Brit., *Cantiaci* (Kent, GB) 51°22′N, 0°12′E
Villa of unique type. Remains of walls, baths, cult rooms, frescoes, mosaics, granary, mausoleum. Romano-Celtic temple. Christian chapel (2nd half 4th c). Museum.
Arch Cant LXX (1956), 249–50; LXXII (1958), xlvii–l: Meates, G. W., *Lullingstone Roman Villa,* London, 1963, and *The Roman Villa at Lullingstone, Kent I,* Maidstone, 1979. Lewis, 1966, 2–3, 111–3: Rivet, 1969, 70, 149.

LUMIGNY Lug., *Meldi*? (Seine-et-Marne, F) 48°44′N, 2°57′E
Gallo-Roman foundations over La Tène dwellings: ?*vicus.*
Gallia XIX (1961), 300.

LUNERAY Lug., *Caletes* (Seine-Maritime, F) 49°50′N, 0°55′E
Gallo-Roman foundations, cremation cemetery.
Cochet, *S.I.,* 289.

LVTETIA *cf* Paris

LUTON *cf* Waulud's Bank

LUTZ-EN-DUNOIS Lug., *Carnutes* (Eure-et-Loir, F) 48°03′N, 1°25′E
Gallo-Roman foundations, tiles, terra sigillata, coins, near the château.
Nouel, 16.

LUZARCHES Lug., *Parisii* (Val-d'Oise, F) 49°07′N, 2°25′E
Gallo-Roman foundations, tiles (?villas).
Duval, 1961, 232: Roblin, 1971, 7, 102, 165, 249.

LYDIARD TREGOZE Brit., *Dobunni* (Wiltshire, GB) 51°33′N, 1°50′W

Pottery; between 10 and 20 kilns, 2nd and 3rd c.
Britannia IV (1973), 317.

LYDNEY Brit., *Dobunni* (Gloucestershire, GB) 51°43′N, 2°33′W

RIB 305–08.

Romano-Celtic temple within an Iron Age fort commanding a stretch of the river Severn. Excavation revealed a timber structure below the existing remains which may represent an Iron Age temple. An iron mine on the site was active in the 3rd and early 4th c. The complex of buildings includes the temple constructed *c* AD 367. Priests' accommodation, a guest-house and baths, and what was probably a row of shops and workshops. Dedicated to Nodons, the basilican temple is one of the most elaborate of its period in the Provinces. The great number of votive objects recovered reveal the god as being concerned with healing, water, the sun and the chase. The temple survived into the 5th c. Reliefs, mosaics, numerous objects. Museum.

Wheeler, R. E. M. and Wheeler, T. V., *Report on the Excavation of the Prehistoric, Roman and Post-Roman Site in Lydney Park, Gloucestershire,* Oxford, 1932: *TBGAS* LXXVIII (1959), 86–91: Ross, 1967, 22, 176, 339 *et al*: Rainey, 1973, 119–21: McWhirr, 1981, 152–5; *Britannia* XIII (1982), 380.

LYMPNE Brit., *Cantiaci* (Kent, GB) 51°04′N, 1°00′E

PORTVS LEMANIS.

It. A. 473.7, 473.10 (*ad portum Lemanis*).

Tab. Peut. (*Lemanio* or *Lemavio*).

Not. Dig. Occ. XXVIII, 5 and 15 (*Praepositus numeri Turnacensium, Lemannis*).

Rav. Cos. 106.35 (*Lemanis*).

PNRB 386–9.

Fort at the mouth of the East Rother (LEMANA FLVMEN: *Rav. Cos.* 108.39). 2nd c, port and probable naval base of the *Classis Britannica* (*RIB* 66); 3rd c fort of the *Litus Saxonicum.* Walls and bastions, bath-house.

VCH Kent III (1932), 55–9: Collingwood-Richmond, 1969, 51: *SxAC* CVII (1969), 102–25: Johnston, 1977, 29–30: Johnson, 1979, 53–6: *Britannia* XI (1980), 227–288.

LYONS-LA-FORÊT Lug., *Veliocasses* (Eure, F) 49°24′N, 1°28′E

Theatre, then probably an agricultural settlement, then cemetery of new-born children; centre of pottery manufacture 1st and 2nd c, site abandoned 3rd c; coins 2nd and 3rd c.

BSNÉP XXXVII, 33–7; XXXIX (1968), 72–8: *BSNAF,* 1967, 250; 1968, 188; 1970, 109: *Ann Norm* XVIII (1968), 283: *Gallia* IX (1951), 84; XXIV (1966), 261; XXVI (1968), 366; XXVIII (1970), 273; XXX (1972), 343.

MACHEN Brit., *Silures* (Gwent, GB) 51°35′N, 3°07′W

Minor settlement, with lead mining, occupied mid 1st to late 2nd c. Floors, columns, pottery of 1st and 2nd c, pieces of lead, coins in lead mine.

Arch Camb XCI (1936), 379; XCIV (1939), 108–10.

MAGIOVINIVM *cf* Dropshort

MAGNY-EN-VEXIN Lug., *Veliocasses* (Val d'Oise, F) 49°09′N, 1°47′E

Minor settlement beside the road between Paris and Rouen, 'La Chaussée Jules César'.

BAVF III (1967), 3, 30.

MAGNY-LA-CAMPAGNE Lug., *Viducasses* (Calvados, F) 49°03′N, 0°06′W

Gallo-Roman foundations of mud and drystone; tiles, painted plaster; stone quarries.

de Caumont, 1831, 233; 1857, 588.

MAGOR FARM *cf* Illogan

MAHALON Lug., *Osismii* (Finistère, F) 48°02′N, 4°26′W

'Lézivy', Gallo-Roman foundations which might be those of a villa.

BSAF, 1875, 132.

MAIDEN CASTLE Brit., *Durotriges* (Dorset, GB) 50°42′N, 2°28′W

4th c. Romano-Celtic temple, within 17-ha. hill fort which was stormed by Vespasian *c* AD 43 and had been deserted by *c* AD 70 in favour of Dorchester. The temple complex comprises a square temple, an oval shrine and a small rectangular building probably the priests' quarters. The oval shrine overlies a pre-Roman timber structure. Finds include a tricorn bull with three female busts on its back.

Wheeler, R. E. M., *Maiden Castle, Dorset,* Oxford, 1943, 131–35: Ross, 1967, 129 *et al*: RCHM *Dorset* II part 3 (1970), 500–01.

MAIDENHATCH FARM *cf* Pangbourne

MAIDSTONE Brit., *Cantiaci* (Kent, GB) 51°16′N, 0°31′E

Major settlement on the road between Rochester and Lympne; foundations, mosaics, cemetery. Museum.

Arch Cant LXXXII (1967), 293: Rivet, 1964, 145: *VCH Kent* III (1932), 98: Rodwell-Rowley, 1975, 63.

MAISONS-EN-BEAUCE Lug., *Carnutes* (Eure-et-Loir, F) 48°24′N, 1°51′E

Gallo-Roman foundations, 'Le Murger', 'La Cave à la Sourde'.

Jalmain, air photograph: Nouel, 20.

MALGUENAC Lug., *Veneti* (Morbihan, F) 48°05′N, 3°03′W

At 'le Guilly', a villa in three parts joined by a corridor 14 m long. At 'Kerhurgan', Gallo-Roman foundations, tegulae.

BSPM, 1899, 137–43; 1901, 336–8: Grenier, *Manuel* II.2 (1934), 873: P-W VIIIA.1 (1955), 774.

MAMERS Lug., *Aulerci Cenomani* (Sarthe, F) 48°21′N, 0°23′E

Temple.

Province du Maine, Oct–Dec 1908, 353; Jan–Mar 1959, 1.

MANCHE (La) *cf* English Channel

MANOIR (Le) Lug., *Baiocasses* (Calvados, F) 49°17′N, 0°36′W
CIL XIII 8976.
Claudian milestone. Foundations, pits. Hoard of 425 coins from Vespasian to Gallienus. 1st to 3rd c.
de Caumont, 1831, 233; 1857, 524–26: *MSAN* XXVIII (1870), 70–72: *BSAN* LI (1948–52), 350.

MANS (LE) Lug., *Cenomani* (Sarthe, F) 48°00′N, 0°12′E
VINDINVM or VINDVNVM, later CENOMANIS.
Ptol. II,8,8 (Οὐίνδινον, Οὐίνδυνον).
Tab. Peut. (*Subdinnum*).
Cf AVLERCI.
Town, capital of the CIVITAS CENOMANORVM. Gallo-Roman fountain; amphitheatre 110 by 80 m, end of 3rd c; walls (3rd c) in the Gaulish style, with towers and posterns, among the best preserved in Gaul. Rectangular grid of streets, three aqueducts, tombs. Museums.
de Caumont, *Abécédaire*, 357: Blanchet, *Enceintes*, 45: *RHAM* LXIII (1908), 92: Ledru, 1911: Grenier, *Manuel* I (1931), 423, 510, 528, 532, 581; III.2 (1958), 840–8: *CRAI*, 1950, 264–5: *Bull Soc agr Sarthe*, 1953–4, 27–45: Cordonnier-Détrie, P., *Le Mans*, Le M., 1954: *Province du Maine*, July–Sept 1956, 161: *JRS* XLVIII (1958), 33–9: Vassas, R., L'enceinte gallo-romaine du M., *Congrès archéologique de France*, 1961: P-W supp IX (1961), 41–9: *RÉA* LXIII (1961), 55: *Gallia* IX (1951), 97; XII (1954), 172; XV (1957), 202; XXIX (1971), 249.

MARAIS (Aux) Bel., *Bellovaci* (Oise, F) 49°26′N, 2°03′E
Pottery kiln, second half of 1st c; local production.
Celticum XV (1965), 225.

MARBOUÉ Lug., *Carnutes* (Eure-et-Loir, F) 48°07′N, 1°20′E
Villa, mosaics; temple, cemetery, 'Mienne'. Gallo-Roman foundations, 'Thuy, Villesard'.
Grenier, *Manuel* II.2 (1934), 838, 842: Nouel, 14, 26.

*****MARCELCAVE** Bel., *Ambiani* (Somme, F) 49°51′N, 2°34′E
Temple, 'Les Tombelles'.
Agache, fig. 212; *BSAP*, 1972, 323: Agache-Bréart, 1975, 87: Agache, 1978, 391, 394.

*****MARCHÉ-ALLOUARDE** Bel., *Ambiani* (Somme, F) 49°44′N, 2°52′E
Gallo-Roman foundations, cellar or sepulchral vault.
BSAP XVI (1888), 396.

MARCILLÉ-RAOUL Lug., *Redones* (Ille-et-Vilaine, F) 48°23′N, 1°36′W
Gallo-Roman foundations, bricks, cement.
BSAIV LIV (1928), 17: Bizeul, *Voie romaine de Rennes vers le Mont Saint-Michel*, 41.

MARCIS
Not. Dig. Occ. XXXVIII, 7 (*equites Dalmatae, Marcis in Litore Saxonico*).
Late Empire fort of Gallia Belgica. Unlocated, but probably in the neighbourhood of Calais (?Marck, ?Marquise, ?Mardyck).
Johnson, 1979, 91–2.

MARCOUSSIS Lug., *Parisii* (Essonne, F) 48°38′N, 2°14′E
Gallo-Roman foundations, ?villa.
Duval, 1961, 233: Roblin, 1971, 8, 184, 300.

MARE AUX PUITS (La) *cf* Rouvray (Foret du)

MARE BRITANNICVM *cf* English Channel

MAREUIL-CAUBERT Bel., *Ambiani* (Somme, F) 50°04′N, 1°50′E
Villa, 'le Domaine'. Pre-Roman *oppidum* of the *Ambiani*, 'Camp de César'.
BSPF XVII (1920), list LXIX, 56: *DAG* (Mt-Caubert): Agache-Bréart, 1975, 88.

MARGAM Brit., *Silures* (West Glamorgan, GB) 51°33′N, 3°42′W
RIB 2255.
Milestone of Postumus (AD 258–68) found 1926 in Coal Brook, Margam; originally presumably beside the road between Caerleon and Neath.
Nash-Williams, 1969, 185 no. 8: Sedgley, 1975, 30.

MARIEUX Bel., *Ambiani* (Somme, F) 50°06′N, 2°26′E
Villa; weapons, ceramics, coins.
Leduque, *Ambianie*, 133: Agache-Bréart, 1975, 88.

*****MARLERS** Bel., *Ambiani* (Somme, F) 49°46′N, 1°51′E
Gallo-Roman foundations, and pre-Roman remains.
Agache, fig. 67: Leduque, *Ambianie*, 163: Vasselle, 323.

MARNE, river *Parisii* at mouth (Seine-et-Marne, F)
MATRONA FLVMEN, later MATERNA.
Caes. *BG* I, 1.
Ausonius *Mosella* 461.
Amm. Marc. XV,11,3.
Sid. Apoll. *Pan. Majorian*. 208.
Iul. Hon. XVI.
Greg. Tur. *HF* V, 29(39); VI, 17(25); VIII, 10.
Rav. Cos. (Maderna).
CIL XIII, 5674.
Duval, 1961, index, *sv*.

MAROEUIL Bel., *Atrebates* (Pas-de-Calais, F) 50°19′N, 2°42′E
Cemetery of the late Empire.
Dérolez, 512, 531.

MARQUISE Bel., *Morini* (Pas-de-Calais, F) 50°49′N, 1°42′E
Stone quarries; altar to Hercules, column to Juno, pottery, coins, iron.
Leduque, *Boulonnais*, 86, n 3; *Morinie*, 69.

MARTINHOE Brit., *Dumnonii* (Devonshire, GB) 51°14′N, 3°55′W
Fortlet; Neronian, evacuated *c* AD 75. Square enclosure containing timber barrack-blocks, etc.,

within outer circular enclosure. Pottery.
PDAS XXIV (1966), 3–39: Branigan-Fowler, 1976, 32–34.

MARTRAGNY Lug., *Baiocasses* (Calvados, F) 49°15′N, 0°36′W
'Ville de Bacai', 'Champ du puits fondu', small villa; bricks, coins.
de Caumont, 1831, 228; 1846, 295.

MATIGNON Lug., *Coriosolites* (Côtes du Nord, F) 48°36′N, 2°18′W
'l'Hôpital', Gallo-Roman foundations; edge of a ?funerary pit.
BSCN XLVII (1909), 43–44.

MATRONA FLVMEN *cf* river Marne

MAULE Lug., ?*Carnutes* (Yvelines, F) 48°55′N, 1°51′E
Cemetery, 4th to 6th c, 'Les Moussets'. Municipal museum.
Gallia XIX (1961), 295; XXIII (1965), 310; XXV (1967), 221; XXVIII (1970), 250.

MAULÉVRIER Lug., *Caletes* (Seine-Maritime, F) 49°33′N, 0°44′E
Villa, corridor, 'Valmont'; 100 m long, basilical (?annexe).
MSAN X (1836–39), 376–87; *Atlas*, 1836: Grenier, *Manuel* II.2 (1934), 802–03: *Arch Comm Ant S.I.*, ms report of excavation.

MAURON Lug., *Veneti* (Morbihan, F) 48°05′N, 2°18′W
At 'Grand-Vaffet' a Gaulish oppidum reoccupied; great numbers of tegulae in proximity to Coriosolite coins. At 'Meslais' Gallo-Roman foundations, hypocaust chamber, tegulae, quern, terra sigillata. At 'Bourg', 'le Clio', 'Lescu', 'Bourrien', Gallo-Roman foundations, tegulae, great numbers of bricks.
BSPM, 1926, 42–43; 1929, 19; 1953, 51: *BSFN*, 1953, 186: *Gallia* XII (1954), 168.

MAYENNE, river Lug., *Aulerci Diablintes* (Mayenne, F)
MEDVANA FLVMEN.
Lucan. I, 438.
Greg. Tur. *HF* X, 9.

MEAUX Lug., *Meldi* (Seine-et-Marne, F) 48°57′N, 2°52′E
IATINVM/IANTINVM or FIXTINNVM/FIXVINVM, later MELDIS.
Ptol. II,8,11 ('Ιάτινον).
Tab. Peut. (*Fixtinnum*).
CIL XIII, 463, 3023–5.
Cf MELDI.
Town, capital of the CIVITAS MELDORVM. Aqueduct, hypocausts, walls. Flourished 2nd to end 3rd c. Musée Bosquet.
Blanchet, *Enceintes*, 82–3 etc: Espérandieu IV (1911), 3207, 3213: Duval, 1961, 224, 285: *BSHADM* V (1964), 354–9; VI (1965), 25–40: *BGASM* X (1969–71), 11: *Arche* XXXIV (May–June 1970), 60: *Gallia* XIX (1961), 300; XXI (1963), 364; XXIII (1965), 318; XXVIII (1970), 243; XXX (1972), 306.

MEDIOLANVM AVLERCORVM (EBVROVICVM) *cf* Évreux, Vieil-Évreux

MEDVANA FLVMEN *cf* river Mayenne

MÉE (Le) Lug., *Senones* (Seine-et-Marne, F) 48°32′N, 2°38′E
Gallo-Roman foundations, 'Le Buisson Pouilleux'; rural settlement late prehistoric – Gallo-Roman.
Gallia XXX (1972), 306.

MELDI
Caes. *BG* V, 5.
Str. IV,3,5 (C.194) (Μέλδοι).
Plin. *NH* IV, 107 (*Meldi liberi*).
Ptol. II,8,11 (Μέλδαι).
Not. Gall. IV, 9 (*civitas Melduorum*).
Ven. Fort. *Carmina* III, 27.
Greg. Tur. *HF* V,1; VII,4; VIII,18; IX,20.
CIL XIII.1, 2924 (*Meldorum*); notes, p 463; ?3024 (*theatrum civi ... M.D.S.P.D.*).
Tribe of Celtic Gaul, included in Lugdunensis (later Lugdunensis Senonia), with capital at IATINVM or FIXTINNVM, later known as MELDIS (Meaux, *qv*).
Desbordes, J.-M., La *civitas Meldorum* et l'ancien diocèse de Meaux, *Bulletin de la Société d'histoire et d'art du diocèse de Meaux* V (1966), 354–9.

MELIN CRYTHAN Brit., *Silures* (West Glamorgan, GB) 51°39′N, 3°49′W
RIB 2257.
Milestone of Diocletian (AD 284–305); on the road between Caerleon and Neath.
Nash-Williams, 1969, 187 no 14: Sedgley, 1975, 31.

MEL(L)E/ODVNVM *cf* Melun

MELUN Lug., *Senones* (Seine-et-Marne, F) 48°32′N, 2°40′E
ME(L)LODVNVM (?METLOSEDVM, METIOSEDVM).
Caes. *BG* VII,58, 60, 61 (*Metiosedum, Metlosedum*).
It. A. 383.2 (*Metledo = Mecletum*?).
Tab. Peut. (*Meteglo = Megledum*?).
Greg. Tur. *Hist. Francorum* VI, 31 (*Meclodonense castrum*).
CIL XIII, 3010–3019.
Major settlement on the road between Sens and Paris, succeeding pre-Roman oppidum of the *Senones*. Museum.
Blanchet, *Enceintes*, 83–5: Espérandieu IV (1911), 99–109: Duval, 1961, 21, 77, 81, 99, 101–5, 114, 223, 224, 236, 251, 290: Nouel, 9: *BGASM* X (1969–71), 25.

MENAPII
Caes. *BG* II, 4; III, 9, 28; IV, 4, 22, 38; VI, 2, 5, 6, 9, 33; VIII, 15.
Str. IV,3,4 and 5 (C.194); IV,5,2 (C.199) (Μενάπιοι).
Plin. *NH* IV, 106 (*Menapi*).
Mart. XIII, 54 (*de Menapis*).
Tac. *Hist.* IV, 28.
Ptol. II,9,5 (Μενάπιοι).
Cass. Dio XXXIX, 44 (Μεναπίους).

Tab. Peut. (*Castello Menapioru(m)*).
Edict. Diocl. IV,1,8 (*petasonis Menapicae*).
Aur. Vic. XXXIX, 20.
Not. Dig. Or. VIII, 3 and 35; *Occ.* V, 75 and 224; VII, 83; XLI, 16.
Oros. I, 2; VI, 7, 8, 10.
Not. Tiron. LXXXVII, 37.
CIL III.2, p 873 = VII, 1195 ((*Cohorte*) *I Menap*(*iorum*)); V, 885 (*cives Menapius*); XI, 390 (*civitatis Menapiorum*); XIII, 624 (*Menap*(*iae*)), 3033 (*de Menapis*); notes, XIII.1, p 567.
Belgic tribe living in the lower Rhine area until 55 BC, when attacks by Germanic tribes forced it to move westward. Under the early Empire the CIVITAS MENAPIORVM had its capital at CASTELLVM MENAPIORVM (Cassel, *qv*); under the late Empire it was replaced by the CIVITAS TVRNACENSIVM, with capital at Tournai (not on this map).
Helinium I (1961), 20–34: *RN* CXCV (Oct–Dec 1967), 721.

MERCATEL Bel., *Atrebates* (Pas-de-Calais, F) 50°14′N, 2°48′E
Villa.
Agache, *BCDMHPC* IX.1 (1971), 49: Agache-Bréart, 1975, 139.

MÉRÉVILLE Lug., *Carnutes* (Essonne, F) 48°19′N, 2°05′E
Three temples, dwellings, Montreau, 'La Remise des Murs, Bois Breton'.
Jalmain, aerial photograph: Duval, 1961, 95.

*MÉRICOURT-L'ABBÉ Bel., *Ambiani* (Somme, F) 49°57′N, 2°34′E
Villa, 'Les Vignes'; temple on a prominent rise in the marsh of Haillon.
Agache, fig. 511; *BSAP,* 1972, 323: Agache-Bréart, 1975, 88.

*MÉRICOURT-SUR-SOMME Bel., *Ambiani* (Somme, F) 49°54′N, 2°40′E
Gallo-Roman foundations; pre-Roman fort, 'des Câteaux', 25 ha.
RN XLIV (1962), 321: Agache-Bréart, 1975, 89.

MÉROUVILLE Lug., *Carnutes* (Eure-et-Loir, F) 48°18′N, 1°54′E
Scattered settlement over 110 ha, 'Sampuy', at the crossroads Sens-Chartres, Blois-Paris; foundations, cellars, wells, statuary, coin hoard, gold and silver vessels.
BSNAF, 1910: *Bull Nat Orl,* July 1959: Nouel, 20, 29, 36: Jalmain, aerial photograph: *Gallia* XXX (1972), 316.

MESLIN Lug., *Coriosolites* (Côtes du Nord, F) 48°27′N, 2°34′W
'la Lande de Gras', Gallo-Roman foundations, building debris; large bronze of Septimius Severus.
MSCN I (1883–84), 235.

MESNIL-ESNARD (Le) Lug., *Veliocasses* (Seine-Maritime, F) 49°25′N, 1°09′E
Villa, ?corridor, 'le Manoir'. Villa, 'La Châtaigneraie'.
Bull Soc émul S.I., 1909, 197–227: *BSNÉP* XVII (1909), 32–38.

MESNULS (Les) Lug., *Carnutes* (Yvelines, F) 48°45′N, 1°50′E
Gallo-Roman foundations, 'La Millière'; hypocaust, plaster.
Gallia XXV (1967), 221; XXVIII (1970), 251.

MESPAUL Lug., *Osismii* (Finistère, F) 48°37′N, 4°01′W
'Bourg', milestone, end of 1st c, on the road between Penzé and Plouescat.
MSHAB, 1956, 19–23: *Gallia* XV (1957), 186.

MIANNAY Bel., *Ambiani* (Somme, F) 50°06′N, 1°43′E
Gallo-Roman foundations, tiles, Merovingian cemetery: foundations, coins, 'St-Pierre-à-Gouy'.
Leduque, *Ambiane,* 62, n 680: Vasselle, 318: Agache-Bréart, 1975, 91: Agache, 1978, 320, 355.

MILDENHALL Brit., *Atrebates* (Wiltshire, GB) 51°25′N, 1°41′W
CVNETIO.
It. A. 486.5 (*Cunetione*).
Rav. Cos. 106.21 (*Cunetzone*).
PNRB 328–9.
Town of 6 ha on the road between Silchester and Bath, at the crossing of the river Kennet by the Cirencester-Winchester road. Occupied throughout the Romano-British period. Earthwork and stone defences on divergent lines. Numerous finds include foundations, tessellated floors and small finds mostly within the bastioned town wall.
WANHM LXI (1966), 9–24: Rodwell-Rowley, 1975, 34–5.

MILLBROOK FARM Brit., *Silures* (Powys, GB) 51°56′N, 3°20′W
RIB 2258, 2259.
Milestone, on the road between Brecon and Abergavenny with inscriptions of Constantius Chlorus (AD 293–306) and Constantine II (AD 317–340).
Nash-Williams, 1969, 187 no 15, 188 no 19: Sedgley, 1975, 31–32 (*S.V.* Llanhamlach).

MINARIACVM *cf* Estaires

MISEREY (Le) Lug., *Aulerci Eburovices* (Eure, F) 49°01′N, 1°16′E
Aqueduct (?of Vieil Evreux).
Ann Norm VIII (1958), 402: *Gallia* XX (1962), 423.

MOINVILLE-LA-JEULIN Lug., *Carnutes* (Eure-et-Loir, F) 48°23′N, 1°42′E
Gallo-Roman foundations, 'Villarsis'.
Jalmain, aerial photograph: Duval, 1961, 42.

*MOLLIENS-VIDAME Bel., *Ambiani* (Somme, F) 49°53′N, 2°01′E
Temple, 2 subterranean rooms with niches, pits, in the woods; foundations, ditches, pits, votive deposits, slag, tiles.
BSAP XIII (1880), 262, 341: Leduque, *Ambianie,* 158 n 591–92: Agache-Bréart, 1975, 95: Agache, 1978, 365.

MONCHY-LE-PREUX Bel., *Atrebates* (Pas-de-Calais, F) 50°16'N, 2°54'E
Villa.
Dérolez, 512: Agache-Bréart, 1975, 140.

MONMOUTH Brit., *Silures* (Gwent, GB) 51°48'N, 2°42'W
?BLESTIVM.
It. A. 485.2.
PNRB 269.
Small settlement, with iron workings, on the road between Gloucester and Usk and at or near the crossing of this road and the road between Caerwent and Wroxeter.
Britannia I (1970), 57, 64–65; V (1974), 400.

MONS-BOUBERT Bel., *Ambiani* (Somme, F) 50°08'N, 1°40'E
Villa.
Agache-Bréart, 1975, 92: Agache, 1978, 68, 253.

MONTABARD Lug., *Esuvii* or *Sagii* (Orne, F) 48°49'N, 0°04'W
Possible fort; square enclosure, reservoirs; a statuette of Venus Anadyomene.
MSAN IX (1835), 457; XIV (1844), 160.

MONTAIGU-LA-BRISETTE Lug., *Unelli* (Manche, F) 49°34'N, 1°24'W
'le Montcastre', Gallo-Roman foundations in Valognes chalk; tiles, bricks, a terra cotta figure; coins of 2nd c.
de Gerville, 1854, 162: *BSAIC* XXIV (1900), 82.

MONTANEL Lug., *Abrincatui* or *Redones* (Manche, F) 48°30'N, 1°25'W
Small villa, burned; coins, pottery.
de Gerville, 1854, 163: *BSAIC* XXIV (1900), 161.

MONTATAIRE Bel., *Bellovaci* (Oise, F) 49°16'N, 2°26'E
Cemetery, Gallo-Roman then Carolingian.
Durvin.

MONTCHATON Lug., *Unelli* (Manche, F) 49°01'N, 1°30'W
Votive deposit, 4th c; a bronze pot and four bronze vessels, arms, eight small late bronze coins.
BSAIC XXIV (1900), 123.

*****MONTDIDIER** Bel., *Ambiani* (Somme, F) 49°39'N, 2°34'E
Villa, probable; small finds; oppidum on the frontier.
Leduque, *Ambianie,* 145 n 371–73: Agache-Bréart, 1975, 92: Agache, 1978, 40, 41, 44, 61, 79, 232, 316.

MONTEREAU Lug., *Senones* (Seine-et-Marne, F) 48°23'N, 2°57'E
CONDATE (SENONVM).
It. A. 383.3.
Tab. Peut.
Major settlement on the road between Paris and Sens, at the confluence of the rivers Seine and Yonne. Piles of bridge (disappeared); coin hoard; at 'La Pièce aux Moines', two villas. Municipal museum.
Desjardins, *Tab. Peut.,* 25: Jalmain, D., *Provins et sa région,* 1967, 32: Nouel, 2: *Rev num,* 6th series, XIV (1972), 184–207.

MONTEREAU-SUR-YARD Lug., *Senones* (Seine-et-Marne, F) 48°30'N, 2°40'E
Cemetery, 3rd c.
Gallia VII (1949), 2.

*****MONTIÈRES-LÈS-AMIENS** Bel., *Ambiani* (Somme, F) 49°55'N, 2°15'E
Large villa.
Agache-Bréart, 1975, 26 (s.v. Amiens): Agache, 1978, 73, 78, 258, 302, 327, 370.

MONTMARTRE Lug., *Parisii* (Ville-de-Paris, F) 48°52'N, 2°20'E
Temple, habitation. *Cf* Paris.
Duval, 1961, 127: Roblin, 1971, 208.

MONTREUIL-SUR-MER Bel., *Morini* (Pas-de-Calais, F) 50°28'N, 1°46'E
LINTOMAGVS.
Tab. Peut.
?LOCVS QVARTENSIS SIVE HORNENSIS (*qv*).
Major settlement on the road between Boulogne and Amiens. Museum. The name should perhaps be read QVINTOMAGVS – 'Market of QVANTIA (Canche)' – *cf* Étaples. The name Montreuil derives from MONASTERIOLVM.
RN XXIII (1937), 260–65; XXX (1948), 184–96: Leduque, *Boulonnais,* 12–13, 37, 38; *Morinie,* 86.

MONT-ST-ÉLOI Bel., *Atrebates* (Pas-de-Calais, F) 50°21'N, 2°42'E
Large villa.
Jelski, 143: Agache-Bréart, 1975, 140: Agache, 1978, 64.

MONT ST MICHEL Lug., *Abrincatui* or *Redones* (Manche, F) 48°38'N, 1°31'W
Gallo-Roman foundations: mosaic, terra sigillata, coins.
BSAIC XXIV (1900), 161.

MONTS D'ERAINES – STE ANNE D'ENTREMONT Lug., *Viducasses* (Calvados, F) 48°56'N, 0°06'W
CIL XIII 3174.
'Château-Tarin', 'les Monts', small villa. Construction of stones without bricks and with thick mortar joints; semicircular apse. Architectural debris, fragment of an altar of Vieux marble.
MSAN XVII (1849), 395–98: *BSAN* XXIX (1913), 225–29.

*****MORCHAIN** Bel., *Ambiani* (Somme, F) 49°47'N, 2°54'E
Gallo-Roman foundations, 'La fosse Chatelain'; two basements, 2nd–3rd c.
Gallia XXIX (1971), 231: Agache-Bréart, 1975, 92.

MORET-SUR-LOING Lug., *Senones* (Seine-et-Marne, F) 48°22'N, 2°49'E
Pottery kiln, Gallo-Roman foundations. Municipal museum.
Jalmain, air photograph: Nouel, 20.

MORIDVNVM DEMETARVM *cf* Carmarthen

MORIEUX Lug., *Coriosolites* (Côtes du Nord, F) 48°31'N, 2°36'W
'La Chapelle Saint-Maurice', Gallo-Roman foundations; debris.
MSCN I (1883–84), 235–36.

MORINI

Caes. *BG* II, 4; III, 9, 28; IV, 21, 22, 37, 38; V, 24; VII, 75, 76.
Vergil. *Aen* VIII, 727.
Str. IV,3,5 (C.194); IV,5,2 (C.199–200) (Μορῖν-οι).
Mela III,23.
Plin. *NH* IV, 102, 106, 122; X, 53; XII, 6; XIX, 8.
Sil. Ital. VII, 605; XV, 723.
Tac. *Hist* IV, 28.
Flor. I, 45.
Ptol. II,9,1 and 4 (Μορινοί).
Cass. Dio XXXIX, 44, 50, 51; LI, 21.
Not. Dig. Occ. XL, 52.
Not. Gall. VI, 12 (*civitas Morinorum*).
Hier. *Epist* CXXIII, 16.
Oros. I, 2; III, 20; VI, 7–9.
Paulinus Nol. *Epist* XVIII, 4.
Beda *Ecc Hist* I, 1 and 2.
CIL III, 2049 (*cohort(is) I Morinorum*); p 864 = VII, 1193 (*cohortibus . . . I Morinorum*); VI, 29692 (*castrensis Morini*); XI, 391 (*cīvitatis Morinorum*); XIII.1, p 560–62 (notes); 3560 ([*civi*]*tas Morinor(um)*); XIII.2, 8727 (*IIvir colon(iae) Morinorum*).
Tribe of Gallia Belgica with capital at TARVENNA (Thérouanne, *qv*). GESORIACVM (Boulogne, *qv*) was also in their *civitas* under the early Empire, but later became a *civitas* in its own right.
BSAAM XIX (1957–8), 1–128: *Helinium* I (1961), 20–39: *Celticum* XV (1966), 151: Delmaire, 1976, *passim*.

MORLAIX Lug., *Osismii* (Finistère, F) 48°35'N, 3°50'W
Vicus; various foundations under the château at the first ford over the Jurlot. Roads towards Carhaix, Coz-Yaudet, Kérilian. Many coins of Gordian III found in the walls of the château.
du Chatellier, 1907, 84.

MORVILLERS-SAINT-SATURNIN Bel., *Ambiani* (Somme, F) 49°47'N, 1°50'E
Major settlement; tiles, mosaics, tesserae, coins from Nero to Tetricus.
Leduque, *Ambianie*, 163, n 701.

MOULINEAUX Lug., *Lexovii* (Seine-Maritime, F) 49°21'N, 0°58'E
Villa, 2nd–3rd c, urns, at 'La Mardole' near the church.
Gallia XX (1962), 427.

MOURS Lug., *Parisii* (Val d'Oise, F) 49°08'N, 2°16'E
Gallo-Roman foundations, mosaics, plaster, coins.
Gallia XXVIII (1970), 250.

MOUTIERS-HUBERT Lug., *Lexovii* (Calvados, F) 48°58'N, 0°16'E
Iron-smelting furnace with slag in the runnel; pottery.
BSAN LII (1952–54), 309–10.

MOYENNEVILLE Bel., *Ambiani* (Somme, F) 50°04'N, 1°45'E
Minor settlement, foundations.
BSAP XII (1875), 183: Vasselle, 323: Agache-Bréart, 1975, 93: Agache, 1978, 422.

MUCH HADHAM Brit., *Catuvellauni* (Hertfordshire, GB) 51°52'N, 0°04'E
Pottery kiln 1st c, four kilns 4th c; two tile kilns.
JRS LIX (1969), 221.

MUCKING Brit., *Trinovantes* (Essex, GB) 51°30'N, 0°26'E
Major settlement, continuous occupation from pre-Roman to Saxon. Cemetery, pottery kiln.
Ant J XLVIII (1968), 213, 215; LIV (1974), 183–99: *Britannia* I (1970), 183; II (1971), 236, 296; III (1972), 334–35; IV (1973), 305; V (1974), 442; VI (1975), 264–65; VII (1976), 344: *JRS* LVII (1967), 190.

MUNTHAM COURT Brit., *Regni* (West Sussex, GB) 50°52'N, 0°25'W
Shrine. Circular floor, all evidence of the superstructure destroyed. Numerous bronze and other votive objects indicated use from late pre-Roman period to 4th c. To south a well with ritual filling. Finds Worthing museum.
ANL V (1954–55), 204–05; VI (1955–60), 101–02.

MUREAUX (Les) Lug., *Carnutes* (Yvelines, F) 49°00'N, 1°55'E
Gallo-Roman foundations.
Gallia XXI (1963), 357.

MUTRÉCY Lug., *Viducasses* (Calvados, F) 49°04'N, 0°25'W
Small villa on the road between Clinchamps and Boulon; foundations.
de Caumont, 1831, 227; 1850, 181.

MUZY Lug., *Aulerci Eburovices* (Eure, F) 48°46'N, 1°21'E
Cemetery, 'La côte Vaubreu', beside the road between Chartres and Évreux.
Gallia XXIV (1966), 240.

MYNHEER FARM Brit., *Dumnonii* (Cornwall, GB) 50°14'N, 5°12'W
RIB 2234.
Milestone of Gordian III (AD 238–44) found in a field about 300 m east of Gwennap Pit.
JRS XXXIV (1944), 88: Sedgley, 1975, 23.

MYNYDD CARN GOCH Brit., *Silures* (West Glamorgan, GB) 51°39'N, 4°01'W
Two practice camps.
JRS LI (1961), 126: Nash-Williams, 1969, 129; RCAHM (Wales), *Glamorgan* I, ii, (1976), 101–5.

MYNYDD MYDDFAI Brit., ?*Silures* (Dyfed, GB) 51°58'N, 3°43'W
Fortlet.
JRS XLVIII (1958), 95–96.

NACQUEVILLE Lug., *Unelli* (Manche, F) 49°40'N, 1°44'W
Gallo-Roman foundations on the shore; a covered cinerary urn, 12 coins. La Tène III foundations and burials.
BSAIC XXIV (1900), 46.

***NAMPS-AU-MONT** Bel., *Ambiani* (Somme, F) 49°49'N, 2°06'E
Villa, 'L'Épinette'.
Agache, fig. 437: Agache-Bréart, 1975, 95: Agache, 1978, 113, 332, 398.

***NAMPS-AU-VAL** Bel., *Ambiani* (Somme, F) 49°48'N, 2°06'E
Three adjacent temples, 'Bois du Fay'.
Agache, fig. 571: Agache-Bréart, 1975, 95: Agache, 1978, 22, 41, 392, 400.

NANSTALLON Brit., *Dumnonii* (Cornwall, GB) 50°28'N, 4°46'W
Fort, built *c*. AD 54, abandoned *c* 80. Area *c* 1 ha; usual internal buildings.
Britannia III (1972), 56–111: Webster, 1981, 63–5, 86.

NANTERRE Lug., *Parisii* (Hauts-de-Seine, F) 48°54'N, 2°12'E
NEMETODVRVM
Greg. Tur. *Hist. Francorum* X, 28 (*Nemptudoro*).
Temple and cult site, late pre-Roman and early Empire; coin hoard, 255–6; chariot burial. Name means 'sacred grove'.
Duval, 1961, 68, 78, 231, 247, 278, 284: Roblin, 1971, 13, 36, 126, 161, 217.

NAOURS Bel., *Ambiani* (Somme, F) 50°02'N, 2°17'E
Villa.
Vasselle, 323: Agache-Bréart, 1975, 95.

NÉANT-SUR-YVEL Lug., *Veneti* (Morbihan, F) 48°01'N, 2°20'W
'la Ville aux Feuves, Feuvres', Gallo-Roman foundations, hypocaust chamber, tegulae, potsherds, glass.
BSPM, 1929, 19.

NEATH Brit., *Silures* (West Glamorgan, GB) 51°40'N, 3°49'W
NIDVM.
It. A. 484.2 (*Nido*).
PNRB 425.
Fort, founded *c* AD 75 and abandoned *c* 130; small settlement or posting station in 3rd c. Walls; pottery of 1st and 2nd c, coins Augustus to Trajan.
Nash-Williams, 1969, 98–101: Webster, 1981, 57.

NEATHAM Brit., ?*Belgae/Regni* (Hampshire, GB) 51°10'N, 0°56'W
Minor settlement, posting station on the road between Silchester and Chichester, established 1st c, most active in 3rd and 4th. Stone buildings, market place, cemetery.
Britannia II (1971), 283; III (1972), 348; IV (1973), 317; VI (1975), 213–6.

NEMETACVM *cf* Arras

NEMETOCENNA *cf* Arras

NEMETODVRVM *cf* Nanterre

NEMETOSTATIO *cf* ?North Tawton

NEMOURS Lug., *Senones* (Seine-et-Marne, F) 48°16'N, 2°42'E
Gallo-Roman foundations, cemetery, statuary. Musée du Vieux-Château.
Nouel, 21, 27, 29.

NESLE Bel., *Ambiani* (Somme, F) 49°46'N, 2°55'E
Villa.
Duhamel-Drujean, *Description du canton de Nesle,* Peronne/St-Quentin, 1884, 49: Vasselle, 323: Agache, fig. 517: Agache-Bréart, 1975, 95.

NETHERWILD FARM *cf* Aldenham

NETTLETON SHRUB Brit., *Belgae/Dobunni* (Wiltshire, GB) 51°29'N, 2°15'W
Possible fort, succeeded by temples.
An original small circular shrine of early Flavian date was associated with a rectangular hall and ?hostel. In 3rd c. shrine was replaced by an octagonal temple on a high podium and the accompanying settlement saw great prosperity. An inscribed altar and bronze plaque indicate a cult of Apollo.
VCH Wiltshire I (1957), 91: *JRS* LII (1962), 191: Lewis, 1966, 158: Ross, 1967, 216: *Britannia* VI (1975), 45: Webster, 1980, 161. Wedlake, W. J., *The Excavation of the Shrine of Apollo at Nettleton, Wiltshire, 1956–71,* London, 1982.

NEUFMONTIERS-LÈS-MEAUX Lug., *Meldi* (Seine-et-Marne, F) 48°59'N, 2°50'E
Gallo-Roman foundations of 4th c, 'Les Touches'.
Gallia XXIII (1965), 318.

NEUILLAC Lug., *Veneti* (Morbihan, F) 48°08'N, 2°59'W
'Porzo', Gallo-Roman foundations within an enclosure 105 m by 62 m in extent surrounded by ditches fed by the river Blavet.
Cayot-Delandre, 1847, 419: Fouquet, 1853, 112: Le Mené, 1891, 39, 43: *BSPM*, 1950, 45–48; 1972, 46: *Ogam* XI (1959), 28.

NEUILLY-SUR-SEINE Lug., *Parisii* (Hauts-de-Seine, F) 48°53'N, 2°16'E
Gallo-Roman foundations, early Empire.
Duval, 1961, 231, 269: Roblin, 1971, 61, 126, 212.

NEUVILLE-LÈS-DIEPPE-BRACQUEMONT Lug., *Caletes* (Seine-Maritime, F) 49°55'N, 1°06'E
Temple, *oppidum*, ?*vicus*, 'Cité de Limes'.
Hardy, M., *Nouvelles recherches sur la cité de L.*, *camp refuge celtique près de Dieppe,* Rouen, 1875: Grenier, *Manuel* II (1934), 232, 753: Wheeler, R. E. M., and Richardson, K., *Hill-forts of Northern France,* Oxford, 1957, 123–5: *Gallia* XXII (1964), 292: *Ogam* XXI (1969), 1–6, 31.

NEUVILLE-SAINT-VAAST Bel., *Atrebates* (Pas-de-Calais, F) 50°21'N, 2°46'E
Kiln.
Jelski, 144.

NEUVILLE-SUR-SARTHE Lug., *Aulerci Cenomani* (Sarthe, F) 48°04'N, 0°12'E

Gallo-Roman foundations, 'La Houdannerie': cemetery, small finds, 'Le Grenouillet'.
Bouton.

NEWPORT Brit., *Belgae* (Isle of Wight, GB) 50°42'N, 1°17'W
Villa, winged corridor. Baths, mosaics probably of 2nd c.
Ant J IX (1929), 141–51: Rainey, 1973, 122.

NEWTON St LOE Brit., *Belgae* (Avon, GB) 51°23'N, 2°25'W
Villa, probably corridor, probably 4th c. Two large buildings, a number of rooms with mosaic floors.
VCH Somerset I (1906), 302: *JRS* XXVI (1936), 43–46: Rainey, 1973, 122–23.

NIBAS Bel., *Ambiani* (Somme, F) 50°06'N, 1°35'E
Cemetery, small finds.
Leduque, *Ambianie,* 165, n. 734.

NIDVM *cf* Neath

NOEUX-LES-MINES Bel., *Atrebates* (Pas-de-Calais, F) 50°29'N, 2°40'E
Kiln.
Dérolez, 516.

NOINTEL *cf* **CLERMONT-DE-L'OISE**

NORDEN Brit., *Durotriges* (Dorset, GB) 50°39'N, 2°04'W
Shrine, well, shale-working. Pottery, bones, miniature altars, coins attesting occupation 1st to 4th c.
RCHM *Dorset* II part 3 (1970), 598.

NORTHCHURCH *cf* Berkhamstead

NORTHFLEET Brit., *Cantiaci* (Kent, GB) 51°27'N, 0°20'E
Lime kiln.
VCH Kent III (1932), 128.

NORTH LEIGH Brit., *Dobunni* (Oxfordshire, GB) 51°50'N, 1°25'W
Large villa, courtyard, developed from 1st-c building and occupied until the 5th c. Mosaics mostly laid in 4th c. Museum. Further unexcavated buildings to S. identified on air photographs.
VCH Oxfordshire I (1939), 316–18: *JRS* XXXIV (1944), 81: Rainey, 1973, 125–6: *Britannia* V (1974), 259; VII (1976), 337; VIII (1977), 400; IX (1978), 444; XI (1980), 372–3.

NORTH SEA
OCEANVS GERMANICVS.
Plin. *NH* IV, 103 (*mare Germanicum*).
Ptol. II,3,4; 11,1; VIII,3,2; 6,2 (Ὠκεανὸς Γερμανικός).
Marcian. II, 31; 44.
PNRB 44.

NORTH TAWTON Brit., *Dumnonii* (Devon, GB) 50°47'N, 3°54'W
?NEMETOSTATIO.
Rav. Cos. 105.47 (*Nemetotacio*).
PNRB 424–5.
Fort, probably built *c* AD 55, at the crossing of the river Taw, beside a road leading westward from Exeter. Area 2.6 ha.
Fox, A., *South-West England,* Newton Abbot, 1973, 164.

NORTH WRAXALL Brit., *Belgae* (Wiltshire, GB) 51°29'N, 2°14'W
Villa; outbuildings, baths, mosaics. Coins from Trajan to Gratian but especially early 4th c.
VCH Wiltshire I (1957), 92.

NOTRE-DAME-DE-FRANQUEVILLE Lug., *Veliocasses* (Seine-Maritime, F) 49°24'N, 1°10'E
Villa.
BCASI XV (1909), 112: *BSNÉP* XVII (1909), 32–8.

NOTRE-DAME-DU-GUILDO Lug., *Coriosolites* (Côtes du Nord, F) 48°35'N, 2°13'W
'Quatre Vaux', large villa.
ACN, 1851, 234–39: Cunat, 1850, 10–14.

NOVIODVNVM DIABLINTVM *cf* Jublains

NOVIOMAGVS *cf* Chichester, Crayford, Lisieux

NOYAL-SOUS-BAZOUGES Lug., *Redones* (Ille-et-Vilaine, F) 48°25'N, 1°38'W
'le Courtil Gaillard', Gallo-Roman foundations including traces of a hypocaust, tegulae.
Gallia XXXIII (1975), 337.

NOYELLES-GODAULT Bel., *Atrebates* (Pas-de-Calais, F) 50°25'N, 2°59'E
Minor settlement, end 1st c, burnt 3rd c, reoccupied 4th c; hypocausts, pottery kilns, coin hoard; ditches, huts, wells, burials. *Cf* Hénin-Liétard.
Gallia XXI (1963), 338; XXIII (1965), 294; XXIX (1971), 230: *RÉA* LXIX (1967), 228–54.

NOYELLES-SUR-MER Bel., *Ambiani* (Somme, F) 50°11'N, 1°43'E
Brick kilns; substantial foundations, villa, hypocausts, small finds.
Prarond VI (1868), 166: Agache, fig. 32–3 *et al:* Vasselle, 323: Leduque, *Ambianie,* 85: Agache-Bréart, 1975, 96.

NOYELLE-VION Bel., *Atrebates* (Pas-de-Calais, F) 50°18'N, 2°33'E
Kiln.
Dérolez, 516.

NVDION(N)VM *cf* ?Sées ?Jublains

NUNNEY Brit., *Belgae* (Somerset, GB) 51°13'N, 2°22'W
Villa, probably courtyard. Baths, mosaics. Probably built *c* AD 300, abandoned *c* 370.
VCH Somerset I (1906), 317: *PSANHS* CXIV (1970), 37–47: Rainey, 1973, 126–27.

NURSLING Brit., *Belgae* (Hampshire, GB) 50°56'N, 1°29'W
Minor settlement; wells, pottery, bronzes, coins AD 70 to 380, mostly after 250.
VCH Hampshire I (1900), 311.

NUTHILLS Brit., *Belgae* (Wiltshire, GB) 51°25'N, 2°03'W

Villa, winged corridor. Painted plaster, coins AD 264–353.
VCH Wiltshire I (1957), 54.

NVTIONNVM *cf* ?Sées ?Jublains

OAKLANDS Brit., *Regni* (East Sussex, GB) 50°55′N, 0°31′E
Iron workings; furnaces, coins, pottery.
Straker, 1931, 329.

OCEANVS BRITANNICVS *cf* English Channel
OCEANVS GERMANICVS *cf* North Sea
OCEANVS HIVERNICVS *cf* Irish Sea
OCRINVM PROMONTORIVM *cf* Lizard (The)
OCTAPITARVM PROMONTORIVM *cf* St David's Head

ODIHAM Brit., *Atrebates* (Hampshire, GB) 51°16′N, 0°56′W
Villa; baths, hypocausts; active in 4th c. Tile kiln.
PHFC X (1926–30), 225–36: *JRS* XXI (1931), 242–43: *Britannia* II (1971), 282–83.

OINVILLE-SOUS-AUNEAU Lug., *Carnutes* (Eure-et-Loir, F) 48°28′N, 1°44′E
Villa.
Jalmain, air photograph: Nouel, 20.

OISE, river Bel./Lug., *Parisii* at mouth (Oise, Val d'Oise, F)
ISARA FLVMEN.
It. A. 384.11 (*Briva Isare*).
Tab. Peut.
CIL XIII, 9158.
Cf Pontoise.

OISON Lug., *Carnutes* (Loiret, F) 48°08′N, 1°58′E
Gallo-Roman foundations.
Gallia XIX (1961), 342.

OISSEAU-LE-PETIT Lug., *Aulerci Cenomani* (Sarthe, F) 48°21′N, 0°05′E
Theatre, settlement on the road between Jublains and Chartres, large square building; abundant terra sigillata, coins of Nero and Valens.
Bouton: *Bull Soc sc. et arts Sarthe* II, 225–39: Pesche, *Dictionnaire* IV, 310–11; V, 481; VI, 720–21: *Province du Maine* I, 286: *RHAM* XXXIV, 115; XXXV, 113; XXXVI, 23; XLVII, 114.

OKEHAMPTON Brit., *Dumnonii* (Devon, GB) 50°44′N, 3°58′W
Fort, area 1 ha, occupied *c* AD 50–80.
Britannia VIII (1977), 415; X (1979), 255–8.

OLD BURROW Brit., *Dumnonii* (Devonshire, GB) 51°14′N, 3°44′W
Fortlet, Neronian, evacuated *c* AD 75. Square enclosure within outer circular enclosure.
PDAS XXIV (1966), 3–39: Branigan-Fowler, 1976, 32–34.

OLDLANDS Brit., *Regni* (East Sussex, GB) 51°01′N, 0°05′E
Iron workings; furnaces, coins of Nero to Diocletian, pottery, skeletons, terra sigillata, glass.
Straker, 1931, 395.

OLD SARUM Brit., *Belgae* (Wiltshire, GB) 51°05′N, 1°48′W
SORVIODVNVM.
It. A. 483.4 (*Sorvioduni*), 486.13 (*Sorbiodoni*).
PNRB 461.
Minor settlement, posting station at important intersection of roads, situated within and around an Iron Age hill fort. Considered possibly to have been the administrative centre of an Imperial estate. Flint and timber buildings, pottery 1st to 4th c, coins mostly 4th c.
WANHM LVI (1955), 102–26: *Britannia* I (1970), 299.

OLINA FLVMEN *cf* river Orne

ORGÈRES-EN-BEAUCE Lug., *Carnutes* (Eure-et-Loir, F) 48°09′N, 1°42′E
Gallo-Roman foundations, 'La Mare de Mai'.
Jalmain, air photograph.

ORGEVILLE *cf* Pacy-sur-Eure

ORIVAL *cf* Rouvray

ORLU Lug., *Carnutes* (Eure-et-Loir, F) 48°22′N, 1°55′E
Villa, coin hoard, 'Le Bis de Maison': Gallo-Roman foundations, 'Bissay, les Cloziots'.
Jalmain, air photograph: Nouel, 20, 31, 36.

ORNE, river Lug., *Viducasses* at mouth (Orne, Calvados, F)
OLINA FLVMEN.
Ptol. II,8,2 ('Ολίνα ποταμοῦ ἐκβολαί).

OROMANSACI (?)
Plin. *NH* IV, 106.
Cited by Pliny between the *Morini* and the *Caletes*; perhaps a client tribe of the *Ambiani* in Ponthieu (Fossier I, 126), but the passage seems corrupt (?read: *Menapi Morini ora Marsacis iuncti pago*). The *Marsaci* lived further east (*cf* Tac. *Hist* VI, 56) and are not shown on this map.

ORPINGTON Brit., *Cantiaci* (Kent, GB) 51°23′N, 0°05′E
Villa, corridor; baths, mosaics.
Arch Cant LXXI (1957), xlvi, 240: *KAR* VIII (1967), 9.

ORSETT Brit., *Trinovantes* (Essex, GB) 51°30′N, 0°22′E
What was formerly regarded as a Roman fort proved on excavation to be a defended farmstead of the late Iron Age and Romano-British periods. The site has accordingly been omitted from the map.
Britannia XI (1980), 35–42.

OSISMII
Caes. *BG* II, 34; III, 9; VII, 75.
Str. I,4,3 and 5 (C. 63–4); IV,4,1 (C. 195) ('Οσίσμιοι, but 'Ωστιμίοι when quoting Pytheas).
Mela III, 23 (*Ossismos*); III, 48 (*Ossismicis*).
Plin. *NH* IV, 107 (*Ossismos*).
Ptol. II,8,5 and 6 ('Οσίσμιοι).
Tab. Peut. (*Osismi*).

Not. Dig. Occ. V, 118 and 268; VII, 94 (*Mauri Osismiaci*); XXXVII, 6 (*Corumosismis*), 17 (*praefectus militum Maurorum Osismiacorum, Osismis*).
Not. Gall. III, 9 (*civitas Ossismorum*).
Steph. Byz. ('Ωστίωνες).
Not. Tiron. LXXXVII, 43 (*Othismus*).
CIL XIII.1, p 490.
Armorican tribe included in Lugdunensis (later Lugdunensis III) occupying Finistère with capital at VORGIVM (Carhaix, *qv*).
Pape, L., *La Civitas des Osismes à l'époque gallo-romaine*, Paris, 1978.

OSPRINGE Brit., *Cantiaci* (Kent, GB) 51°18′N, 0°51′E
?DVROLEVVM.
It. A. 472.4 (*Durolevo*).
Tab. Peut. (*Durolevo*).
?*Rav. Cos.* 108.37 (*Durolavi*, included in the section on rivers).
PNRB 351.
Major settlement on the road between London and Canterbury; altars, mausolea, cemetery, remains of walls, mosaics.
Whiting, W., Hawley, W., and May, T., *Report on the Excavation of the Roman Cemetery at Ospringe, Kent*, London and Oxford, 1931: *VCH Kent* III (1932), 93–6: *Ant J* XXXI (1951), 38 (*sub* Faversham); XLIX (1969), 253–94: *KAR* VII (1967), 20; IX (1967), 6; XIV (1968), 2: *Arch Cant* LXXXIII (1968), 260: *Britannia* I (1970), 44, 73.

OTFORD Brit., *Cantiaci* (Kent, GB) 51°18′N, 0°11′E
Cemetery; villa.
Arch Cant XLII (1930), 157–71; LXXXI (1961), lxi: *JRS* LIX (1969), 232: *KAR* VI (1966), 8–13; XIII (1968), 4–5: Rivet, 1969, 91, 145: *VCH Kent* III (1932), 122.

OTHONA *cf* Bradwell-on-Sea

OUARVILLE Lug., *Carnutes* (Eure-et-Loir, F) 48°24′N, 1°46′E
Two villas, 'La Remise St-Philippe', 'Les Perrières'.
Jalmain, air photographs.

OUISTREHAM Lug., *Viducasses* (Calvados, F) 49°17′N, 0°15′W
'Delle du Castelier', 'Castillon', fort beside the Orne on the road between Bayeux and Bénouville. One gold coin of Lucius Verus, 2nd c. Burials in stone sarcophagi oriented west–east.
de Caumont, 1846, 424–28: *BSAN* VI (1874), 308–09.

OXFORD Brit., ?*Catuvellauni* (Oxfordshire, GB) 51°45′N, 1°16′W
Minor settlement. Numerous small finds, no significant structures but pottery kilns in vicinity. Occupation pre-Roman, Romano-British and post-Roman.
VCH Oxfordshire I (1939), 301–03.

OZOIR-LE-BREUIL Lug., *Carnutes* (Eure-et-Loir, F) 48°01′N, 1°29′E
Gallo-Roman foundations.
Nouel, 20.

PACY-SUR-EURE Lug., *Aulerci Eburovices* (Eure, F) 49°01′N, 1°23′E
Temple, ?villa, 'Orneville'.
BSNÉP XV (1907), 65–75.

PAGANS HILL Brit., *Belgae* (Avon, GB) 51°22′N, 2°38′W
Romano-Celtic temple; octagonal, with central *cella* and surrounding ambulatory. Ranges of rectangular buildings on the north and east; well on the west. Built late 3rd c, in use until early 5th c.
PSANHS XCVI (1951), 112–42; CI (1956–57), 15–51: Lewis, 1966, *passim*.

*PAILLART Bel., *Ambiani* (Oise, F) 49°40′N, 2°19′E
Two villas; sarcophagus.
Agache, fig. 235: Leduque, *Ambianie*, 150: Agache-Bréart, 1975, 149: Agache, 1978, 138, 245, 272, 381, 412.

PAIMPOL Lug., *Osismii* (Côtes du Nord, F) 48°46′N, 3°03′W
Gallo-Roman foundations; coins, a golden bracelet weighing two ounces.
MSCN I (1883–84), 209.

PAINSWICK Brit., *Dobunni* (Gloucestershire, GB) 51°48′N, 2°12′W
Villa, courtyard. Probably built late 2nd c. Mosaic.
TBGAS XXVII (1904), 156–71.

PALEY Lug., *Senones* (Seine-et-Marne, F) 48°14′N, 2°51′E
Cemetery; villa.
Jalmain, air photograph: Nouel, 27.

PAMBER Brit., *Atrebates* (Hampshire, GB) 51°20′N, 1°06′W
Tile kiln, one fragment stamped with name of Nero.
Ant J VI (1926), 75–76; *Britannia* XII (1981), 290–1.

PANGBOURNE Brit., *Atrebates* (Berkshire, GB) 51°28′N, 1°07′W
Villa, corridor, at Maidenhatch Farm. Built *c* AD 200; aisled outbuilding built early 4th c.
Britannia II (1971), 284.

PANNES Lug., *Senones* (Loiret, F) 48°01′N, 2°40′E
Gallo-Roman foundations; cemetery; coin hoard.
Nouel, 21, 27, 35.

PARAMÉ Lug., *Coriosolites* (Ille-et-Vilaine, F) 48°40′N, 2°00′W
'Boulevard de Rochebonne', Gallo-Roman foundations, tegulae, potsherds etc.
BSAIV VII (1868), CV: *Ann Bret* LX (1953), 217: Langouet, 1973, 148.

PARIS Lug., *Parisii* (Ville-de-Paris, F) 48°52′N, 2°20′E
LVTETIA PARISIORVM, later PARISIIS.
Caes. *BG* VI, 3; VII, 57, 58 (*Lutecia/Lutetia/Lucetia/Lucecia*).
Str. IV,3,5 (C.194) (Λουκοτοκίαν).
Ptol. II,8,10 (Λευκοτεκία).
It. A. 366.5, 368.2, 383.1, 384.1,7,8,12 (*Luticia*).

Tab. Peut. (Luteci).
Julian. Misopogon 340 (Λουκετίαν).
Amm. Marc. XV,11,3 (Parisiorum castellum, Luticia nomine).
Vib. Seq. (Lutecia).
Not. Tiron. LXXXVII, 10 (Lutecia).
CIL XIII.1, p 464.
Cf PARISII.

Town, capital of the CIVITAS PARISIORVM. Oppidum established c 250 BC on the Île de la Cité in the river Seine (wooden bridges). The name Lutetia has as its root a word meaning 'mud'. Gallo-Roman town established under the early Empire on the hill Sainte-Geneviève; regular plan, the main axis being the road to Orléans. Forum on the Rue Soufflot, enlarged at the beginning of the 2nd c to 160 × 100 m; temple on the west side, basilica on the east. Theatre, early 2nd c, on the Rue Saint-Louis (external diameter 72 m, seats for 4500). Amphitheatre, with stage, on the Rue Monge (greatest length 138 m, arena 58 × 52 m, stage 40 m). Possible circus at the Halle des Vins. Three baths: 'Thermes de l'Est' (Collège de France), 'Thermes du Forum' (Rue Gay-Lussac) and 'Thermes du Nord' (Cluny, 'Palais de Julien'; late 2nd–early 3rd c, façade marked by the Boulevard Saint-Germain, 100 × 65 m from north to south; 250 m² of the frigidarium surviving, with vaults resting on consoles ornamented by prows of ships). Aqueduct of Arceuil-Cachan, 15 km long. At Montmartre (Mons Mercurii, NOT Martyrum) temple of Mercury (columns re-used in the church of St-Pierre); possible temple of Mars, houses. Dwelling with hypocaust at 1 Rue Gay-Lussac; hypocausts in the Rue P.-Nicole and along the Rue de la Tombe-Issoire; necropolis at the Luxembourg; Christian cemetery at the Gobelins. Sculptures: monument to boatmen (the nautae Parisiaci), column-base showing eight gods, Jupiter column, monuments with military decoration, architectural elements re-used in the late-Empire enclosure.

Signs of a conflagration a little after the middle of the 3rd c. The Île de la Cité, the original nucleus of Lutetia, thereafter served as a refuge (enclosing wall rebuilt c 275). Here Julian was acclaimed Emperor by his troops in 360. Evidence of Christianity from the second half of the 3rd c; pilgrimage of St Marcel, 4th c. At Saint-Denis (qv) large Christian cemetery developed round the tombs of SS Denis, Rusticus and Eleutherus.

Museums: Cabinet des Medailles (Bibliothèque Nationale), Musée Carnavalet, Louvre, Musée des Thermes et de l'Hôtel de Cluny.

de Pachtère, F., Paris à l'époque gallo-romaine, P., 1912: Grenier, Manuel I (1931), 416; III.1 (1958), 363–7; III.2 (1958), 815–8, 899–903; IV.1 (1960), 180–91, 310–19: BSNAF, 1950–51, 189: Dion, R., Le site et la croissance de Paris, Revue des Deux Mondes I.1 (1951): RÉA LIII (1951), 301–11; 1962, 425 and 428: Paris et Île-de-France, Mém. VII (1956), 41; XV (1964), 7: Beaujeu, J., La géographie de Lutèce, Geographia, April 1957, 7; May, 8: Duval, P-M., Les inscriptions antiques de Paris, I-II, 1960; Duval, 1961: Art de France II (1962), 229:

Durand-Lefebvre, Marques de potiers gallo-romains trouvées a P., P., 1963: RH, Jan–March 1963, 41: RIO XVII (1965), 275–288: JS, Jan–March 1968, 5: Actes du VIIIe Congrès G. Budé, P., 1968, 54–66: Roblin, 1971: Gallia XVII (1959), 267–72; XXIV (1966), 235–8; XXV (1967), 205–12; XXVIII (1970), 239; XXX (1972), 301; XXXIII (1975), 319–24; XXXV (1977), 321–5; XXXVII (1979), 333–4.

PARISII
Caes. BG VI, 3; VII, 4, 34, 57, 75.
Str. IV,3,5 (C.194) (Παρίσιοι).
Plin. NH IV, 107 (Parisi).
Ptol. II,8,10 (Παρίσιοι).
It. A. 366.5 (Luticia Parisiorum).
Tab. Peut. (Parisi).
Conc. Agrip. a. 346.
Julian. Misopogon 340.
Conc. Paris. a. 360.
Amm. Marc. XV,11; XVII,2, 8; XVIII,6; XX,1, 4, 5, 8, 9; XXI,2; XXVII,2, 10.
Sulp. Sev. Vita S Martini XVIII, 3.
Not. Dig. Occ. XLII, 23 and 66.
Not. Gall. IV, 8 (civitas Parisiorum).
Zosimus III, 9, 1.
Vib. Seq.
Conc. Aurel. a. 511.
Ven. Fort., Greg. Tur., etc., passim.
CIL XIII.1, 626 (civis Parisius); 2924; p 464 (notes); 3026 (nautae Parisiaci); 3034 (Pari-[siorum]); XIII.2, 8974 (milestone, a civ(itate) Par(isiorum)).

Tribe of Celtic Gaul included in Lugdunensis (later Lugdunensis Senonia), with capital at LVTETIA PARISIORVM, later known as PARISIIS (Paris, qv).

Duval, 1961: Colbert de Beaulieu, J.-B., Les monnaies gauloises des P., Paris, 1970: MEFR LXXXII (1970), 15–41: P-W Supp. XII (1970), article Parisii: Roblin, 1971.

PARK FARM cf Aylburton

PARK STREET Brit., Catuvellauni (Hertfordshire, GB) 51°43′N, 0°20′W
Villa; timber house built c AD 65 on site of native farmstead enlarged in masonry c 150 and again late 3rd c; abandoned mid-4th c. Detached bath house and other outbuildings. Tile kiln.
Arch J CII (1945), 21–110; CXVIII (1961), 100–35: Herts Arch II (1970), 62–65: Todd, 1978, 33–58, passim.

PELVES Bel., Atrebates (Pas-de-Calais, F) 50°17′N, 2°55′E
Gallo-Roman foundations.
Jelski, 140.

PENVÉNAN Lug., Osismii (Côtes du Nord, F) 48°50′N, 3°22′W
'Castel Bras', Gallo-Roman foundations; large, medium and small bronze coins of the early Empire. At 'Pors-Guen' building debris.
MSCN I (1883–84), 359.

PEN-Y-COEDCAE Brit., *Silures* (Mid Glamorgan, GB) 51°35′N, 3°21′W
Temporary camp.
JRS XLIX (1959), 102; *RCAHM (Wales), Glamorgan* I, ii, 1976, 99, 102.

PEN-Y-DARREN Brit., *Silures* (Mid Glamorgan, GB) 51°45′N, 3°22′W
Fort, occupied AD 74 to *c* 120. Bath buildings.
Nash-Williams, 1969, 106–08; *RCAHM (Wales), Glamorgan* I, ii, 1976, 84–6.

PEN-Y-GAER Brit., *Silures* (Powys, GB) 51°53′N, 3°12′W
Fort; occupied *c* AD 75 to 120.
Nash-Williams, 1969, 108–10.

PEPPERING-EYE Brit., *Regni* (East Sussex, GB) 50°53′N, 0°28′E
Iron workings, furnaces, terra sigillata.
Straker, 1931, 351.

PERCY-EN-AUGE Lug., *Viducasses* (Calvados, F) 49°03′N, 0°03′W
Cemetery of burials with arms in sarcophagi oriented northwest–southeast beside the road between Vendoeuvre and Lisieux. A small, slender vessel glazed brilliant black.
de Caumont, 1867, 492: *BSAN* XXXV (1924), 475–78.

PERROS-GUIREC Lug., *Osismii* (Côtes du Nord, F) 48°49′N, 3°27′W
'Ar Hastel', Gallo-Roman foundations.
MSCN I (1883–84), 297.

PERTHES-EN-GÂTINAIS Lug., *Senones* (Seine-et-Marne, F) 48°29′N, 2°34′E
Gallo-Roman foundations, hypocaust, plaster.
Gallia XXVIII (1970), 245.

PETERSTOW Brit., *Dobunni* (Herefordshire, GB) 51°55′N, 2°38′W
Iron working; forges, slag, great quantities of cinders. Coin of Philip.
VCH Herefordshire I (1908), 171, 193.

PETIT-COURONNE Lug., *Veliocasses* (Seine-Maritime, F) 49°23′N, 1°01′E
Villa; coin hoard, Tetricus to Probus.
Gallia XX (1962), 429; XXIV (1966), 272.

PETROMANTALVM *cf* ?Guiry-en-Vexin

PEVENSEY Brit., *Regni* (East Sussex, GB) 50°49′N, 0°19′E
ANDERITVM.
Not. Dig. Occ. XXVIII, 10 and 20 (*Praepositus numeri Abulcorum, Anderidos*).
Rav. Cos. 106.33 (*Anderelio*).
PNRB 250–2.
Fort of the *Litus Saxonicum*, founded *c* 340. Taken by the Saxons AD 491 (*Anglo-Saxon Chronicle*). Walls and bastions; museum.
Saltzman, L. F., *Excavations on the site of the Roman fortress at Pevensey*, London, 1907; *Excavations at Pevensey*, London, 1908: *VCH Sussex* III (1935), 5–6: *JRS* XXXVIII (1948), 54: Peers, C., *Pevensey Castle, Sussex*, London, 1960:

Cunliffe, 1968, 258, 265, 271: Collingwood-Richmond, 1969, 51: *SxAC* CVII (1969), 102–25: Cunliffe, 1973, 38–9: Johnston, 1977, *passim*: Johnson, 1979, 56–60.

PHRVDIS FLVMEN *cf* river Somme

PICKFORD MILL Brit., *Catuvellauni* (Hertfordshire, GB) 51°50′N, 0°21′W
Barrow. Described when opened 1827 or 1829 as 16 m diameter, 6 m high. Contained circular stone cist with lid; inside a glass vessel and four Drag. 33 samian cups now in British Museum.
VCH Hertfordshire IV (1914), 153.

*PICQUIGNY Bel., *Ambiani* (Somme, F) 49°57′N, 2°09′E
Rural buildings and enclosures; Roman road.
Agache, fig. 49, 58, 138, 236, 251, 262, 296, 564: *Gallia* XXI (1963), 344: Leduque, *Ambianie*, 74, n 327–28: Agache-Bréart, 1975, 101: Agache, 1978, *passim*.

PIERREFONDS Bel., *Suessiones* (Oise, F) 49°21′N, 2°59′E
Minor settlement, 'ville des Gaules'; from Trajan to 4th c, including two destructions. Gallo-Roman foundations, 'Mont-Berny', dwellings, cellar with niche, pits, pool, hypocaust, temple, baths.
RN XLIX (1967), 691: *Gallia* XXV (1967), 199; XXVII (1969), 234.

*PIERREGOT Bel., *Ambiani* (Somme, F) 50°00′N, 2°23′E
Cemetery; Roman road bridge, small finds.
Agache, fig. 18: Leduque, *Ambianie*, 133: Agache-Bréart, 1975, 101: Agache, 1978, 245.

PIERRELAYE Lug., *Parisii* (Val d'Oise, F) 49°01′N, 2°10′E
PETRA LATA.
Cemetery, early Empire; coins of Tiberius, Nero, Domitian.
Duval, 1961, 233, 242: Roblin, 1971, 122, 191.

*PIERREPONT-SUR-AVRE Bel., *Ambiani* (Somme, F) 49°43′N, 2°33′E
Gallo-Roman foundations; cemetery, small finds.
Agache, fig. 135: Leduque, *Ambianie*, 145: Vasselle, 325: Agache-Bréart, 1975, 101.

PIPPINGFORD Brit., *Regni* (East Sussex, GB) 51°03′N, 0°07′E
Iron workings; furnace, pottery.
SASN II (1971), 3.

PISSY-PÔVILLE Lug., *Caletes* (Seine-Maritime, F) 49°32′N, 1°00′E
Villa, near the presbytery.
Cochet, *S.I.*, 188.

PITLANDS FARM *cf* Up Marden

PITMEADS Brit., *Belgae* (Wiltshire, GB) 51°11′N, 2°08′W
Villa; baths, mosaics; coin of Claudius II.
VCH Wiltshire I (1957), 110: Rainey, 1973, 128.

PITNEY Brit., *Durotriges* (Somerset, GB) 51°04′N, 2°47′W

Villa; courtyard. Flourished 3rd and 4th c. Baths, mosaics.
VCH Somerset I (1906), 326–27: Rainey, 1973, 129.

PÎTRES Lug., *Veliocasses* (Eure, F) 49°19'N, 1°14'E
PISTAS (660 –).
Minor settlement, two bath-buildings, theatre, aqueduct, pits, cellar with niche, enclosure.
Bull Mon, 1864, 105: *MSAN* XXIV (1859), 156–65, 398–402.

PLANCOËT Lug., *Coriosolites* (Côtes du Nord, F) 48°32'N, 2°15'W
'le Grand-Trait', villa; debris.
MSCN I (1883–84), 71.

PLANGUENOAL Lug., *Coriosolites* (Côtes du Nord, F) 48°32'N, 2°35'W
Extensive Gallo-Roman foundations near Predero.
MSCN I (1883–84), 244.

PLASNES Lug., *Lexovii* (Eure, F) 49°08'N, 0°37'E
Villa.
Coutil III (*Bernay*), 77: Grenier, *Manuel* II.2 (1934), 870.

PLAXTOL Brit., *Cantiaci* (Kent, GB) 51°15'N, 0°18'E
Villa, inscribed tiles. Tumulus, enclosed within walls. Skeleton (?female), fibulae, glass flagon, bronze fittings, wooden casket, quern, pottery, terra sigillata, tiles.
VCH Kent III (1932), 163, *Britannia* II (1971), 297–8.

PLÉHÉREL Lug., *Coriosolites* (Côtes du Nord, F) 48°38'N, 2°22'W
'Sables d'Or les Pins', villa; debris; quadruple inhumation; coin of Gallienus: 4th to 5th c.
Gallia XXXIII (1975), 364–5.

PLÉLO Lug., *Osismii* (Côtes du Nord, F) 48°34'N, 2°57'W
'Ville-Balin', villa. Debris.
MSCN I (1883–84), 180.

PLÉNÉE-JUGON Lug., *Coriosolites* (Côtes du Nord, F) 48°22'N, 2°24'W
At 'la Ville Josse' a villa, near the road between Corseul and Vannes. At 'la Mare Pilais', 'le Haut Temple' and 'la Kevée' Gallo-Roman foundations, those at 'le Haut Temple' extending over nearly one hectare. Debris, coins.
MSCN I (1883–84), 426: *BSCN* XLVII (1909), 35.

PLÉNEUF Lug., *Coriosolites* (Côtes du Nord, F) 48°36'N, 2°33'W
At 'Dahouet', Gallo-Roman foundations.
BSCN XLVII (1909).

PLÉRIN Lug., *Osismii* (Côtes du Nord, F) 48°32'N, 2°47'W
At 'Port-Aurel' and at 'Peignard, camp des Nonais', Gallo-Roman foundations, building debris; at the former coins of Tetricus etc.

Beside the road between Coz-Yeaudet and Cesson, in a brook near 'les Boisseries', about 30 coins of Titus, Trajan, Hadrian, Lucius Verus, Commodus, etc.
MSCN I (1883–84), 146–47.

PLERNEUF Lug., *Osismii* (Côtes du Nord, F) 48°31'N, 2°53'W
Near 'la Belleissue' Gallo-Roman foundations, debris.
MSCN I (1883–84), 183.

PLÉSIDY Lug., *Osismii* (Côtes du Nord, F) 48°27'N, 3°07'W
Gallo-Roman foundations, building debris, a statuette and a vase in bronze; two querns, coins of Vespasian.
MSCN I (1883–84), 45.

PLESLIN Lug., *Coriosolites* (Côtes du Nord, F) 48°32'N, 2°04'W
'Carna', villa; hypocaust, walls, tegulae.
BSCN I (1883–84), 81.

*PLESSIER-ROZAINVILLERS Bel., *Ambiani* (Somme, F) 49°45'N, 2°33'E
Gallo-Roman foundations, coins.
Vasselle, 326: Agache, 1978, 422.

PLESSIS-GASSOT (Le) Lug., *Parisii* (Val-d'Oise, F) 49°02'N, 2°25'E
Villa, 2nd to 4th c, 'Les Thuileaux'.
BAVF VI (1970), 111.

PLESSIS-GRIMOULT Lug., *Viducasses* (Calvados, F) 48°57'N, 0°36'W
Gallo-Roman foundations. Plated coins of Postumus, Tetricus, Gallienus, 3rd c.
de Caumont, 1857, 230: *BSAN* XXIX (1913), 228.

PLESTAN Lug., *Coriosolites* (Côtes du Nord, F) 48°25'N, 2°27'W
At 'le Chaussix', Gallo-Roman foundations, tegulae, head of a stone statue, small bronze coins of the late Empire.
BSCN XLVII (1909), 37: *MSCN* I (1883–84), 538.

PLEUDIHEN Lug., *Coriosolites* (Côtes du Nord, F) 48°31'N, 1°58'W
At 'la Motte Pilandelle' Gallo-Roman foundations.
BSCN V (1867), appendix 20; XLVII (1909), 20: Guennou, 1965, *Pleudihen*: Langouet, 1973, 152.

PLÉVENON Lug., *Coriosolites* (Côtes du Nord, F) 48°40'N, 2°20'W
'Anse de la Saudraye', large villa; mosaics.
ACN, 1837, 147: *BSCN* XLVII (1909), 49.

PLOENEVEZ PORZAY Lug., *Osismii* (Finistère, F) 48°06'N, 4°12'W
'Treguer', villa; small finds.
du Chatellier, 1907, 174: *Gallia* XXX (1972), 217.

PLOENEVEZ PORZAY Lug., *Osismii* (Finistère, F) 48°06'N, 4°13'W
At 'Camezen', 'Kervel' and 'Trefeuntec', salt manufactories on the Bay of Douarnenez. Various small finds of 1st to 4th c.

du Chatellier, 1907, 173–74: *BSAF*, 1969, 36–37: *Gallia* XXX (1972), 215–16.

PLOGONNEC Lug., *Osismii* (Finistère, F) 48°05′N, 4°11′W
'Lezodoaré', Gallo-Roman foundations which might be those of a villa. Granite head.
BSAF, 1931; 1973, 79–81.

PLOMODIERN Lug., *Osismii* (Finistère, F) 48°11′N, 4°14′W
At 'Goulit-ar-Guer', Gallo-Roman foundations which might be those of a villa, near the road between Douarnenez and the Crozon peninsula; terra sigillata 2nd c. At 'Landrein' Gallo-Roman foundations; statuette in white clay.
BSAF, 1875, 136: *Gallia* XXX (1972), 213.

PLOUBALAY Lug., *Coriosolites* (Côtes du Nord, F) 48°35′N, 2°09′W
At 'la Ville-Bague', Gallo-Roman foundations, building debris, six querns.
BSCN V (1867), appendix 60; XLVII (1909), 82.

PLOUBEZRE Lug., *Osismii* (Côtes du Nord, F) 48°42′N, 3°27′W
At 'Ancienne forteresse de Runéfao', Gallo-Roman foundations, bricks, edged tile, cement.
MSCN I (1883–84), 277–78.

PLOUDALMÉZEAU Lug., *Osismii* (Finistère, F) 48°32′N, 4°39′W
'Ridiny', Gallo-Roman foundations; terra sigillata from the end of the 2nd c.
BSAF, 1972, 80–81.

PLOUDANIEL Lug., *Osismii* (Finistère, F) 48°32′N, 4°19′W
At 'Lanignez' a villa; fragments of bronze; terra sigillata of the end of the 2nd c. Another villa; an iron bit; very abundant coarse pottery, terra sigillata and metallescent pottery from the beginning of the 2nd to the 3rd c. This villa not yet published.
BSAF, 1968, 16.

PLOUESCAT Lug., *Osismii* (Finistère, F) 48°40′N, 4°10′W
'Gorré-Bloué', a villa; baths. Coins of Constantine; pottery of the 2nd c, terra sigillata (Argonne) of the 4th c. Fragments of a glass bracelet, etc.
BSAF, 1916, pl I.

PLOUGRESCANT Lug., *Osismii* (Côtes du Nord, F) 48°51′N, 3°14′W
At 'la Pointe de Chastel' Gallo-Roman foundations bonded with very resistant mortar.
MSCN I (1883–84), 360.

PLOUHINEC Lug., *Osismii* (Finistère, F) 48°01′N, 4°29′W
'Poulgoazec', Gallo-Roman foundations; coarse pottery and terra sigillata; 2nd c.
du Chatellier, 1907, 303.

PLOUJEAN Lug., *Osismii* (Finistère, F) 48°36′N, 3°50′W
'La Boissière', Gallo-Roman foundations. Golden pendant of 'pseudo-Angerona' type.
BSAF, 1857, 138.

PLOUNEVENTER Lug., *Osismii* (Finistère, F) 48°31′N, 4°13′W
'Kérilien-Coatalec', town. Many foundations, theatre; bronze smithy; incinerations; very abundant local and imported pottery, terra sigillata, amphorae, etc. Several metal objects; coins from the beginning of the 1st c to Honorius.
Ann Bret, 1963–1969.

PLOUZANÉ Lug., *Osismii* (Finistère, F) 48°23′N, 4°37′W
At 'le Cosquier' = 'the old house', a villa; abundant coarse pottery of the 2nd c, terra sigillata of the 2nd and 3rd c. At 'Langongar', a tile kiln, near the road between Saint Renan and the sea.
Arm 2/9/1845: *BSAF*, 1969, 38–40; 1973, 61–63.

*POIX Bel., *Ambiani* (Somme, F) 49°47′N, 1°59′E
Gallo-Roman foundations, small finds.
Fossier, 194: Leduque, *Ambianie*, 156: Agache, 1978, 40, 41.

POLIGNY Lug., *Senones* (Seine-et-Marne, F) 48°13′N, 2°45′E
Gallo-Roman foundations, coin hoard.
Nouel, 21, 36.

PONCHES-ESTRUVAL Bel., *Ambiani*/?*Morini* (Somme, F) 50°19′N, 1°54′E
PONTES.
It. A. 363.1 (*Pontibus*).
CIL XIII.2, p 684.
Major settlement, 'Le Moulinel', on the road between Amiens and Boulogne, at the crossing of the river Authie.
Desjardins, *Géo. Gaule* IV, 51: Miller, 1916, 68, 142: *BSNAF*, 1931, 82–4: P-W XXII.1 (1953), 18–19: Leduque, *Ambianie*, 128.

PONDRON Bel., ?*Silvanectes* (Oise, F) 49°16′N, 2°57′E
?RATOMAGVS (*qv*).
RODONVM (Carolingian period).
Minor settlement.
BSNAF, 1879, 91.

PONT-CROIX Lug., *Osismii* (Finistère, F) 48°02′N, 4°29′W
'Kervennec', a villa: rich, with separate bathhouse; mosaics, *opus spicatum*. Various bronze objects – ornaments, harness trappings etc.: fragments of glass vessels. Coins from the 2nd c to about 360. Abundant coarse pottery, terra sigillata of the 2nd c, Argonne terra sigillata very plentiful.
BSAF, 1971, 56–63; 1972, 87–96; 1973, 64–68.

PONT-DE-GENNES Lug., *Aulerci Cenomani* (Sarthe, F) 48°03′N, 0°26′E
Villa, 'Genneau'.
Bouton.

PONT-DE-METZ Bel., *Ambiani* (Somme, F) 49°53′N, 2°15′E
Cemetery, small finds.
Agache, fig. 54: Leduque, *Ambianie*, 155 seq: Vasselle, 326: Agache-Bréart, 1975, 102: Agache, 1978, 38, 159, 160.

PONTES *cf* Ponches-Estruval, Staines

PONTHOU (Le) Lug., *Osismii* (Finistère, F) 48°34′N, 3°38′W

'Ar C'hastel', Gallo-Roman foundations which might be those of a villa on the road between Morlaix and Guingamp.

du Chatellier, 1907, 93.

PONTOISE Lug., *Parisii* (Val-d'Oise, F) 49°03′N, 2°06′E

BRIVA (later PONS) ISARAE.

It. A. 384.11 (*Briva Isare*).

Tab. Peut. (*Bruvsara*).

CIL XIII, 2, p 682–84 (milestone of Tongres).

Major settlement on road, 'Chaussée Jules César'. The name indicates the existence of a bridge which the topography suggests and which is also indicated by pieces of timber discovered in the course of draining about 1875–80; coins.

Dutilleux, A., *Recherches sur les routes anciennes du département de Seine-et-Oise*, Versailles, 1881, 24–5: Desjardins, *Géog. Gaule* IV (1893), 62, 137: Duval, 1961, 77, 114, 241.

PONT-RÉMY Bel., *Ambiani* (Somme, F) 50°03′N, 1°55′E

Villa.

Agache, fig. 254, 281, 529: *BSÉA* X (1918), 319–27: Agache-Bréart, 1975, 102: Agache, 1978, 219, 227, 262, 315, 338, 368.

PONT-SAINTE-MAXENCE Bel., *Silvanectes* (Oise, F) 49°18′N, 2°36′E

Minor settlement; bronze hermaphrodite, coin hoard.

Roblin, M., L'habitat ancien dans la région de P., *Mélanges Piganiol*, Paris, 1966, 1087–1110; Roblin, 1978, 225–9.

POPE'S HILL Brit., *Dobunni* (Gloucestershire, GB) 51°49′N, 2°28′W

Iron working; furnace, slag, pottery 2nd c to late 4th c.

TBGAS LXXV (1956), 199–202.

PORDIC Lug., *Osismii* (Côtes du Nord, F) 48°35′N, 2°49′W

Gallo-Roman foundations on the point of L'Ermo; fragments of bronze swords; pot containing coins of Probus and his successors.

MSCN I (1883–84), 152–53.

PORT-BAIL Lug., *Unelli* (Manche, F) 49°20′N, 1°41′W

'Goucy-St Marc', 'la Mangreve Jennetot', 'Lanquetot'; *vicus* or small settlement, with a port; on roads between hamlet of St Marc, Valognes and Carentan. Various foundations, aqueducts, baths, hypocausts. Coins of the 3rd and 4th c, coin hoard of the 4th c. Slabs of marble from Vieux. A votive deposit, possibly indicating a temple, was found near the source of Lanquetot; coins, a bronze amulet, statuettes in terra cotta.

MSAN V (1829–30), lxii, lxiii, 24–32: *BSAIC* XXIV (1900), 78–88: *BSNÉP* XIII (1905), 154–6: *BSAN* XXXV (1923), 542–5; LII (1952–4), 283; LIII (1955–6), 197–263: *Ann 5 Norm*, 1926, 10–30, 49–55; 1953, 49, 57.

PORTCHESTER Brit., *Regni* (Hampshire, GB) 50°50′N, 1°07′W

?PORTVS ADVRNI or ARDAONI (*qv*).

Fort of the *Litus Saxonicum*, built late 3rd c ?under Carausius. Occupation (probably by German mercenaries) continued as late as early 5th c. Walls and bastions.

Cunliffe, B., *Excavations at Portchester Castle* I, London, 1975; II, London, 1976: Johnson, 1979, 60–63.

PORT-LE-GRAND Bel., *Ambiani* (Somme, F) 50°09′N, 1°45′E

Large villa, associated with an indigenous farm; urns, coins, small finds.

MSÉA, 1878, 304–5: Agache, fig. 22, 37–9, 116, 123, 195, 204, 655: Vasselle, 326: Leduque, *Ambianie*, 162 n 669–70: Agache-Bréart, 1975, 102–3.

PORT TALBOT Brit., *Silures* (West Glamorgan, GB) 51°35′N, 3°48′W; 51°34′N, 3°45′W

RIB 2252–2256.

Milestones, on the road between Caerleon and Neath with inscriptions of Gordian (AD 238–44), Diocletian (282–305) and Licinius (308–24).

Nash-Williams, 1969, 184 no 5, 186 no 13, 187 no 17, 188 no 18: Sedgley, 1975, 30–31.

PORTVS AEPATIACI

Not. Dig. Occ. XXXVIII, 5 and 9 (*Tribunus militum Nerviorum, Portu Aepatiaci*).

Location uncertain: Oudenburg, Étaples, Le Tréport, Isques, Boulogne (as a corruption of *Portus Gesoriacus*).

BSAAM, 1880, 113: Grenier, *Manuel* I (1931), 391: *CRAI*, 1944, 372–86: *RÉA* XLVI (1944), 299–317: Leduque, *Boulonnais*, 56: Johnson, 1979, 93.

PORTVS ARDAONI

Not. Dig. Occ. XXVIII, 11 and 21 (*Praepositus numeri exploratorum, Portum Adurni*).

?*Rav. Cos.* 106.19 (*Ardaoneon*).

PNRB 441–2.

Fort of the *Litus Saxonicum*. It is disputed whether it should be identified with Portchester or with Walton Castle, probably the former.

PORTVS DVBRIS *cf* Dover

PORTVS ITIVS *cf* ?Wissant

PORTVS LEMANIS *cf* Lympne

PORTVS VLTERIOR *cf* Tréport, Le

POTBRIDGE FARM *cf* Odiham

POTTERS BAR Brit., *Catuvellauni* (Hertfordshire, GB) 51°42′N, 0°11′W

Tile kiln, 1st c.

CBA Group 10 *Newsletter* VI, July 1961, 12a.

POULLAN Lug., *Osismii* (Finistère, F) 48°05′N, 4°25′W

'Kerandraon', salt manufactory on the shore of the Bay of Douarnenez. One coin; coarse pottery.

du Chatellier, 1907, 264: *Gallia* XXX (1972), 217.

POUPRY Lug., *Carnutes* (Eure-et-Loir, F) 48°06′N, 1°50′E
Gallo-Roman foundations.
La France illustrée, 1852: *Rev Soc arch Orl* IV, 156.

PRÉFONTAINE Lug., *Senones*? (Loiret, F) 48°06′N, 2°41′E
Cemetery.
Nouel, 27, 28.

PRESTON Brit., *Cantiaci* (Kent, GB) 51°17′N, 1°12′E
Pottery kiln.
VCH Kent III (1932), 131.

PRESTON Brit., *Regni* (East Sussex, GB) 50°50′N, 0°08′W
Villa, corridor; mosaics. Built 2nd c, destroyed late 3rd c.
VCH Sussex III (1935), 23–24.

PRIDDY Brit., *Belgae* (Somerset, GB) 51°15′N, 2°39′W
Lead working. Coins, samian ware.
VCH Somerset I (1906), 335: *Britannia* III (1972), 344.

PRIEST WOOD *cf* Cromhall

PRIMROSE HILL Brit., *Durotriges* (Dorset, GB) 50°37′N, 2°01′W
Stone quarry, Purbeck Marble.
RCHM *Dorset* II part 3 (1970), 620.

PRITTLEWELL Brit., *Trinovantes* (Essex, GB) 51°33′N, 0°42′E
Minor settlement; cemetery, foundations.
Rivet, 1964, 147: *VCH Essex* III (1963), 167.

*PROYART Bel., *Ambiani* (Somme, F) 49°53′N, 2°42′E
Temple, 'Le fief de Bac, Petite Vallée'.
Agache, fig. 213, 470, 485; *BSAP*, 1972, 323: Agache-Bréart, 1975, 104: Agache, 1978, 293, 301, 392, 393, 395, 396, 402.

PUCKERIDGE Brit., *Trinovantes* (Hertfordshire, GB) 51°53′N, 0°01′E
1st c. settlement, 2nd c. cemetery. Partridge, C. *Skeleton Green*, London, 1981. See also Braughing.

PUISEUX Bel., *Atrebates* (Pas-de-Calais, F) 50°07′N, 2°42′E
Temple, 'L'Arbre de Serre'.
Agache, 1972, 323: Agache-Bréart, 1975, 140: Agache, 1978, 400.

PULBOROUGH Brit., *Regni* (West Sussex, GB) 50°58′N, 0°29′W
Villa at Borough Farm; baths, mosaics, coins of 1st c, terra sigillata mould fragments; mausoleum at Homestreet Farm, to south.
VCH Sussex III (1935), 25, 63: Cunliffe, 1973, 77–8. Hull 1963, 46.

PURBECK *cf* Primrose Hill

PYLE Brit., *Silures* (Mid Glamorgan, GB) 51°32′N, 3°42′W
RIB 2251.

Milestone, of the road between Caerleon and Neath; Victorinus (AD 268–70).
Nash-Williams, 1969, 186 no 10: Sedgley, 1975, 30.

QVARTENSIS LOCVS *cf* ?Saint-Valéry-sur-Somme

QUEBRIAC Lug., *Redones* (Ille-et-Vilaine, F) 48°21′N, 1°50′W
At 'route de Tinteniac', Gallo-Roman foundations; others elsewhere, with coins.
Orain, 1882, 106: *Bull Arch Ass Bret*, 1887, 231: *BSAIV* LIV (1927–29), 152.

QUEMENEVEN Lug., *Osismii* (Finistère, F) 48°07′N, 4°07′W
Villa, bathhouse; on the road between Quimper and Châteaulin.
BSAF, 1875, 140.

QUEND Bel., *Ambiani* (Somme, F) 50°19′N, 1°38′E
Gallo-Roman foundations, small finds, hoard.
Dufetelle, A., *Monographie de Quend*, Abbeville, 1907: *Gallia* XXV (1967), 204: Leduque, *Ambianie*, 28: Vasselle, 326: Agache, 1978, 37, 247.

*QUERRIEU Bel., *Ambiani* (Somme, F) 49°56′N, 2°25′E
Villa, within pre-Roman enclosure; coins, sarcophagi.
Vasselle, 326: Agache-Bréart, 1975, 106: Agache, 1978, 178, 245, 370.

QUINTIN Lug., *Osismii* (Côtes du Nord, F) 48°24′N, 2°55′W
At 'Roquiniac', Gallo-Roman foundations.
MSCN I (1883–84), 262.

QUIOU (Le) Lug., *Coriosolites* (Côtes du Nord, F) 48°21′N, 2°00′W
In the town Gallo-Roman foundations, building debris.
BSCN XLVII (1909), 31.

*QUIRY-LE-SEC Bel., *Ambiani* (Somme, F) 49°40′N, 2°23′E
Fort, probable.
Agache, fig. 185, 301, 614: *Celticum* XV (1966), 139–50: Agache, 1978, 232.

RADEPONT Lug., *Veliocasses* (Eure, F) 49°21′N, 1°20′E
Possible site of RITVMAGVS, qv.

RADLETT Brit., *Catuvellauni* (Hertfordshire, GB) 51°41′N, 0°19′W
Pottery kilns; two of potter CASTVS.
VCH Hertfordshire IV (1914), 159–60.

RAINVILLERS Bel., *Bellovaci* (Oise, F) 49°25′N, 2°00′E
Gallo-Roman foundations, mosaics, 'Les Orgeries'.
Gallia Sup. X.1.1 (1957), no 83.

RATOMAGVS
1) Alternative name for ROTOMAGVS (Rouen, *qv*).
2) Settlement of the *Silvanectes*.
Ptol. II,9,6 ('Ρατόμαγος).
CIL XIII, 3475 (VIC(O) RATVM[agensiu]M?). Found at Hermes.
Identification disputed: *cf* Hermes, Pondron, Senlis.
Longnon, A., Conjectures sur l'emplacement de Ratomagus, chef-lieu des Silvanectes, *BSNAF*, 1879, 91: Matherat, G., Ratomagus Silvanectum, sa position, *Hommages à M. Renard* III, 418–430.

RECULVER Brit., *Cantiaci* (Kent, GB) 51°22'N, 1°10'E
REGVLBIVM.
Not. Dig. Occ. XXVIII, 8 (*Regulbi*), 18 (*Tribunus cohortis primae Baetasiorum, Regulbio*).
PNRB 446–7.
Fort at the mouth of the Wantsum channel separating Thanet from the Kentish mainland. Pre-Roman occupation; small fort, Claudian period; fort of the *Litus Saxonicum*, 3rd and 4th c. *Vicus*, funerary pits, inscriptions.
Arch Cant LXXXIII (1969), 296: *BMQ* XXIV (1969), 58–63: Collingwood-Richmond, 1969, 49: *JRS* LIX (1969), 233, 242: *KAR* XVII (1969), 18–20: *Britannia* I (1970), 304: Philp, B., *The Roman Fort at Reculver*, 6th edn, Dover, 1970: Johnston, 1977, 15: Frere, 1978, 211–2: Johnson, 1979, 45–8.

REDONES or **RIEDONES**
Caes. *BG* II, 34; VII, 75 (*Rhedones* or *Redones*).
Plin. *NH* IV, 107 (*Rhedones*).
Ptol. II,8,9 ('Ρήδονες).
Not. Dig. Occ. XLII, 36 (*Redonas*).
Not. Gall. III, 4 (*civitas Redonum*).
Not. Tiron. LXXXVII, 68 n 90 Z (*Redonas*).
Conc. Aurel. a. 511 (*episcopus de Redonis*).
Ven. Fort. *Vita S. Paterni* X.
Greg. Tur. *HF* V, 22, 24; VIII, 32, 42; IX, 24; X, 9.
CIL XIII, 3152 ([*civ*]*itas Ried*[*onum*]); 8953–5, 8958, 8959, 8961, 8970 (milestones, C(*ivitate*) R(*edonum*)).
Armorican tribe included in Lugdunensis (later Lugdunensis III) with capital at CONDATE REDONVM, later known as REDONAS (Rennes, *qv*).
Colbert de Beaulieu I, 123–5.

REGINEA *cf* ?Erquy

REGNI (?REGINI, ?REGNENSES)
Ptol. II,3,13 ('Ρῆγνοι).
It. A. 477.10 (*a Regno*).
Rav. Cos. 106.20 (*Navimago Regentium*).
PNRB 445–6.
British *civitas* or tribe, occupying Sussex and southern Surrey. Of uncertain status in the pre-conquest period (most of the coins in the region are of the *Atrebates*). Included in the client kingdom of Cogidubnus, established as a *civitas* in the Flavian period, with capital at NOVIOMAGVS (Chichester, *qv*).
Rivet, 1964, 158–9: Cunliffe, 1973, *passim*: Wacher, 1975, 239–245: *Britannia* I (1970), 78–9; X (1979), 227–254: Salway, 1981, 748–52.

REGVLBIVM *cf* Reculver

REILLY Lug., *Veliocasses* (Oise, F) 49°15'N, 1°51'E
Gallo-Roman foundations, 13 units, 1st to 4th c; Roman roads.
BAVF IV (1968), 85–94.

*REMIENCOURT Bel., *Ambiani* (Somme, F) 49°47'N, 2°23'E
Temple; ditches and enclosure, cellars, foundations, 'Bois du Bucail'. Villa.
BSAP XXVI (1914), 3: Vasselle, 327: Agache, fig. 210, 211; *Id.* 1972, 323: Agache-Bréart, 1975, 109.

*RENANCOURT-LÈS-AMIENS Bel., *Ambiani* (Somme, F) 49°54'N, 2°17'E
Gallo-Roman remains, villa, hypocaust, ?trenches.
Vasselle, 327: Agache, 166.

RENNES Lug., *Redones* (Ille-et-Vilaine, F) 48°05'N, 1°41'W
CONDATE REDONVM, later REDONAS.
Ptol. II,8,9 (Κονδάτε).
It. A. 386.6, 387.3.
Tab. Peut.
CIL XIII, 3148–53.
Cf REDONES.
Town, capital of the CIVITAS REDONVM; road junction at the confluence of the rivers Ille and Vilaine. Established in the Augustan period, at its greatest 2nd c, ended early 5th c. Town walls with multiple battlements; cemetery with cremations and inhumations; Augustan pottery (mortaria); many industrial remains; many thousand coins from Augustus to Valentinian III; a golden patera.
Gallia XXIX (1971), 109–22: Pape, L., Rennes Antique, in *Histoire de R.*, Jean Meyer, Toulouse, 1972, 27–64 (succinct bibliography, p 8).

RETTENDON Brit., *Trinovantes* (Essex, GB) 51°38'N, 0°32'E
Two pottery kilns, 4th c.
JRS LVIII (1968), 197.

REVELLES Bel., *Ambiani* (Somme, F) 49°51'N, 2°07'E
Temples, one larger, 'Bois de la Vallée'.
BSAP XXI (1903), 615: Agache, fig. 237, 408, 436, 574, 630, 631; 1972, 323: Vasselle, 327: Agache-Bréart, 1975, 109: Agache, 1978, 397, 398, 400.

RHUS *cf* Epiais-Rhus

*RIBEMONT-SUR-ANCRE Bel., *Ambiani* (Somme, F) 49°57'N, 2°34'E
Temple, Romano-Celtic, on Iron Age site. Temenos enclosed by Romans with wall and portico, with a classical vestibule on the facade. Situated on the crest of a hill, oriented towards the sunrise. Built under Augustus or Tiberius, destroyed *c* 160; rebuilt early 3rd c, destroyed in the invasions, 2nd half of

3rd c. The third version was a modest structure, built by the local inhabitants on the site of the vestibule and much used. At the back is a vast landscape crossed by the Roman road between Amiens and Bavai; to the east a sanctuary on the gentle slope down to the river. Opposite the temple, two large courts, bordered laterally by two series of sizeable structures about 800 m long. Baths towards the valley, on the axis of the temple; between the two a theatre several times remodelled. Architectural fragments, capitals, sculpture, inscriptions.

Agache, fig. 202, 399, 400, 444, 577, 581: Agache-Bréart, 1975, 110–11 (plan): *BSAP*, 1862, 97; 1912, 318–23; 1971, 43–70; 1972, 323: *Gallia* XXV (1967), 202; XXVII (1969), 228; XXIX (1971), 231; XL (1982), 83–105: Leduque, *Ambianie,* 135: *RN* XLVIII (1966), 539–43: no 207 (Oct–Dec 1970), 469–511.

RICHBOROUGH Brit., *Cantiaci* (Kent, GB) 51°17′N, 1°18′E
RVTVPIAE.
Lucan. *Pharsalia* VI, 67 (*Rutupina litora*).
Juv. IV, 141 (*Rutupino fundo*).
Ptol. II,3,12 ('Ρουτουπίαι).
It. A. 463.4, 466.5, 472.6 (*ad portum Ritupis*).
It. M. 496.4 (*ad portum Ritupium*).
Tab. Peut. (*Ratupis*).
Amm. Marc. XX,1,3; XXVII,8,6 (*ad Rutupias*).
Ausonius *Parentalia* VII, 2; XVIII, 8; *Ordo Urb. Nob.* VII, 9 (*Rutupina, -us, -um*).
Not. Dig. Occ. XXVIII, 9 and 19 (*Praefectus legionis secundae Augustae, Rutupis*).
Oros. I,2,76 (*Rutupi portus*).
Rav. Cos. 106.36 (*Rutupis*).
PNRB 448–450.
(*Rutupinus* was used in verse as an alternative to *Britannicus*).

Fort, town and port at the mouth of the Wantsum channel separating Thanet from the mainland. Bridgehead of Claudius, AD 43; military depot and naval station, 44–85; great monument, of which the foundations still survive, constructed c 85–100; port and commercial town, 2nd c; small fort (c 0.2 ha) built c 250, with earthen rampart and triple ditch; fort of the *Litus Saxonicum* (2.4 ha) built c 275–285, with massive wall and double ditch. Lupicinus landed here in 360 and Theodosius in 368. Walls, amphitheatre, workshops, baths, cemetery, funerary pits, two Romano-Celtic temples, Christian church, lime kiln. Museum.

Smith, C. R., *The Antiquities of Richborough, Reculver and Lymne, in Kent,* London, 1850: Bushe-Fox, J. P., *Excavations at the Roman Fort at Richborough,* Oxford, I (1926); II (1928); III (1932); IV (1949): Ross, 1967, 133–5, 381; 1968, 271: Cunliffe, 1968, *passim*: Collingwood-Richmond, 1969, 49: *SxAC* CVII (1969), 102–125: *Britannia* I (1970), 78, 182, 240–8; II (1971), 225–31: Johnston, 1977, *passim*; Johnson, 1979, 48–51: Webster, 1980, 95–6 etc.; *RIB* 46–65.

RICHVILLE *cf* ?*Ritumagus*
RIDVNA INSVLA *cf* Alderney/Île d'Aurigny
RIGOIALENSIS VICVS *cf* Rueil
RITVMAGVS
It. A. 382.4.
Tab. Peut.
CIL XIII, 2, p 682.
Town of the Veliocasses. Situation uncertain: ?Radepont (Eure), Fleury/Andelle, Charleval, Richville.

RIVENHALL Brit., *Trinovantes* (Essex, GB) 51°49′N, 0°39′E
Villa, courtyard; frescoes, hypocaust.
VCH Essex III (1963), 171–4: *Britannia* IV (1973), 115–27: *Ant J* LII (1973), 219–31: Todd, 1978, 15–18.

ROCHE-DERRIEN (La) Lug., *Osismii* (Côtes du Nord, F) 48°45′N, 3°16′W
At 'le Bouret', at a point between la Roche-Derrien and le Pont de Treguier and at another location, Gallo-Roman foundations and debris; coins of Postumus and Marius.
MSCN I (1883–84), 343–44.

ROCHE-MAURICE (La) Lug., *Osismii* (Finistère, F) 48°28′N, 4°12′W
'Valy-Cloistre', villa with separate baths. Fragments of metal, abundant potsherds both local and imported. Sesterces of the 2nd c, a denarius of Hadrian, an aureus of Vespasian.
Ann Bret LXXIX (1972), 215–51.

ROCHESTER Brit., *Cantiaci* (Kent, GB) 51°23′N, 0°30′E
DVROBRIVAE.
It. A. 472.3 (*Durobrovis*); 473.3 (*Dubobrius*); 473.8 (*Durobrivis*).
Tab. Peut. (. . . *roribus*).
Rav. Cos. 106.37 (*Durobrabis*).
JRS L (1960), 108–11 (. . . *brivas*).
PNRB 346–8.
Town on the road between London and Canterbury, at the crossing of the river Medway. Pre-Roman oppidum with mint, possibly near the scene of the decisive battle in AD 43 between the British forces and the Roman invading army under Aulus Plautius (Cass. Dio LX, 20: *cf* Webster, 1980, 200–202). ?Roman fort, succeeded by town with walls, cemetery, bridge across the Medway. Museum.
VCH Kent III (1932), 80: *History* XXXIX (1953), 105–15: *Arch Cant* LXXVI (1961)–LXXXV (1970): Rivet, 1964, 85, 145, 166: Dudley-Webster, 1965, 67–70: Wacher, 1966, 62, 105, 109: *Britannia* I (1970), 73, 183, 304: Frere, 1978, 80–81.

ROCK Brit., *Belgae* (Isle of Wight, GB) 50°39′N, 1°24′W
Villa; coins of 3rd and 4th c.
VCH Hampshire I (1900), 318: *Britannia* VII (1976), 367–9.

ROCKBOURNE Brit., *Durotriges* (Hampshire, GB) 50°57′N, 1°50′W
Villa, courtyard; baths, mosaics, small finds. Occupied 1st to 4th c. Two stones, ?milestones, inscribed to Decius and Tetricus.
Rainey, 1973, 130–31: Morley Hewitt, A. T., *The*

Roman Villa at West Park, Rockbourne, 1974: *JRS* LII (1962), 195; LVI (1966), 219: Sedgley, 1975, 21.

*RODIVM *cf* Roiglise

RODMARTON Brit., *Dobunni* (Gloucestershire, GB) 51°41′N, 2°05′W
Villa, corridor. Mosaics; date unestablished.
Arch XVIII (1817), 113–16.

*ROGY Bel., *Ambiani* (Somme, F) 49°42′N, 2°13′E
Villa.
Vasselle, 328: Agache-Bréart, 1975, 113: Agache, 1978, 295, 323, 352, 379.

*ROIGLISE Bel., *Ambiani* (Somme, F) 49°41′N, 2°50′E
RODIVM.
Tab. Peut.
ROVDIVM.
Milestone of Tongres, *CIL* XIII, 9158.
Major settlement on the road between Amiens and Soissons. Cemetery, coins, weapons: villa.
Leduque, *Ambianie*, 144: Agache-Bréart, 1975, 113: Agache, 1978, 423.

*ROLLOT Bel., *Ambiani* (Somme, F) 49°35′N, 2°39′E
Gallo-Roman foundations, hoard, coins.
Vasselle, 328.

ROMESCAMPS Bel., *Ambiani* (Oise, F) 49°43′N, 1°48′E
Fort, probable; sarcophagus.
Leduque, *Ambianie*, 157.

ROMFORD Brit., *Trinovantes* (Essex, GB) 51°35′N, 0°11′E
?DVROLITVM (but *cf* Chigwell).
It. A. 480.7 (*Durolito*).
PNRB 352.
Major settlement on the road between London and Colchester. Various remains, centre not yet located.
TEAS VII (1900), 95: *VCH Essex* III (1963), 175: *Britannia* I (1970), 52, 73.

ROMILLY-SUR-ANDELLE Lug., *Veliocasses* (Eure, F) 49°20′N, 1°16′E
Pottery kiln.
Gallia IX (1951), 84.

ROSAY-SUR-LIEURE Lug., *Veliocasses* (Eure, F) 49°22′N, 1°26′E
Gallo-Roman foundations, hypocaust.
Gallia XXIV (1966), 262.

ROSCOFF Lug., *Osismii* (Finistère, F) 48°44′N, 3°59′W
'Laber', Gallo-Roman foundations. Stone statue; terra sigillata of 1st and 2nd c.
Rev Arm, 1844, 345–63.

ROTHAMSTED Brit., *Catuvellauni* (Hertfordshire, GB) 51°49′N, 0°23′W
Tomb, circular, 3.05 m in internal diameter within masonry wall, at centre of a walled enclosure with external ditch, measuring internally *c* 35 m square in which were subsidiary burials, pottery of 1st and 2nd c including some pre-Roman Belgic in a ditch partly overlain by the tomb. Fragments of a small stone statue.
TStAHAAS ns V (1936–38), 108–14.

ROTOIALENSIS VICVS *cf* Rueil

ROTOMAGVS *cf* Rouen

ROTTOMAGVS *cf* Ruan

ROVDIVM *cf* Roiglise

ROUEN Lug., *Veliocasses* (Seine-Maritime, F) 49°26′N, 1°05′E
RATOMAGVS, later ROTOMAGVS
Coins of the *Veliocasses* (*Ratumacos*).
Ptol. II,8,7 ('Ρατόμαγος).
It. A. 382.3, 384.1 (*Ratomago*).
Tab. Peut. (Ratomago).
Conc. Arel. a. 314 (*de civitate Rotomagensium*).
Conc. Agrip. a. 346 (*Rothomagensium*).
Amm. Marc. XV, 41, 12 (*Rotomagi*).
Not. Dig. Occ. XXXVII, 10 and 21 (*praefectus militum Ursariensium, Rotomago*).
Not. Gall. II, 2 (*metropolis civitas Rotomagensium*).
Not. Tiron. LXXXVII, 57 (*Rotomagus*).
Paulinus Nol. *epist* XVIII, 5 (*Ratomagum*).
Conc. Aurel. a. 511.
Ven. Fort.
Greg. Tur.
CIL XIII. 1, p 512.
Important town on Iron Age site; capital under the early Empire of the CIVITAS VELIOCASSIVM, under the later Empire of the CIVITAS ROTOMAGENSIVM (including *Veliocasses* and *Caletes*), and also of the province *Lugdunensis Secunda*. Metropolitan see from end of 4th c. Regular street plan; foundations in the region of the Palais de Justice and the Place du Vieux-Marché, hypocausts, painted plaster, mosaics, remains of monumental inscriptions, sculptures, small finds, two cremation cemeteries. New development at end of 3rd c: enclosing walls, amphitheatre, three inhumation cemeteries. Military base of the *Tractus Armoricanus* (*Not. Dig., supra*). Musée des Antiquités de la Seine-Maritime (catalogues: Vernier, 1923; Flavigny, R.C., 1954; Verron, G., 1971): Musée le Secq des Tournelles.
Desjardins, *Géo Gaule* II, 462; III, 492: Blanchet, *Enceintes*, 33–5: Grenier, *Manuel* I (1931), 422: *Actes du 91e Congrès national des Sociétés savantes, Rouen-Caen, 1956* (Paris, 1958) 23: *RSSHN* XL (1965), 9–16; LI (1968), 15–25: *Ann Norm* XVI (1966), 333–41: Sennequier, G., Rouen gallo-romain et mérovingien, *Connaître Rouen*, 1970 (bibliography): Johnson, 1979, 83–4: *Gallia* VII (1949), 236; XVII (1959), 336; XX (1962), 429; XXIV (1966), 272; XXXII (1974), 333; XXXIV (1976), 336; XXXVI (1978), 310; XXXVIII (1980), 354–63.

ROUMARE Lug., *Veliocasses* (Seine-Maritime, F) 49°24′N, 0°59′E.
Temple, Forêt de Roumare.
BCASI XI (1897), 204: Vesly, 21.

ROUVRAY (Forêt du) Lug., *Veliocasses* (Seine-Maritime, F) 49°21′N, 1°02′E
Temples at Orival, les Essarts, la Mare du Puits. Minor agricultural settlement. 'Le Catelier', oppidum of the *Veliocasses*. *Cf* Grand-Couronne.
BSNÉP X (1902), 139; XI (1903), 131: Vesly, 78–113: Deglatigny, 1927, 7, 13: Grenier, *Manuel* II.2 (1934), 767–70.

ROUVROY Bel., *Atrebates* (Pas-de-Calais, F) 50°23′N, 2°54′E
Villa.
Dérolez, 512.

ROUVROY-LÈS-MERLES Bel., *Ambiani/ Bellovaci* (Oise, F) 49°39′N, 2°21′E
Sanctuary on the border between the *Ambiani* and the *Bellovaci*, in the neighbourhood of several Roman roads. Traces of foundations and levelled terraces over 12 ha of the plateau of 'Mont-Catillon' which overlooks the source of the Rouvoye. At the south of the site, temple with enclosure; on the north-east, theatre and baths. Coin hoard of Claudius. To the north, Roman fortress at Folleville (*qv*).
Cambry, 1803.2, 336: Graves, 1856: *CR et Mém Soc Clermont*, 1948, V: *Gallia* VII (1949), 114; XII (1954), 145; XVII (1959), 285: *Bull Soc arch Creil* XX (1958), 4–8; XXI (1958), 3–7: Agache, fig. 193; *BSAP*, 1972, 323: Agache-Bréart, 1975, 150.

*ROYE Bel., *Ambiani* (Somme, F) 49°42′N, 2°48′E
Roman fort, road; *vicus*, 'Vieux-Catil'.
Grenier, *Manuel* I (1931), 257: Vasselle, 328: Agache, fig. 482: Leduque, *Ambianie*, 144: Agache-Bréart, 1975, 113.

ROZ COUESNON Lug., *Redones* (Ille-et-Vilaine, F) 48°35′N, 1°36′W
'la Rue', Gallo-Roman foundations, *tegulae*.
Gallia XXXIII (1975), 341.

RUAN Lug., *Carnutes* (Loiret, F) 48°07′N, 1°56′E
RVANVM, ROTTOMAGVS, RVTTENV (13th c).
Minor settlement; foundations, statuary, terra sigillata, coin hoard.
Nouel, 10, 20, 29, 35.

RVANVM *cf* Ruan

RUBEMPRÉ Bel., *Ambiani* (Somme, F) 50°01′N, 2°23′E
Villa.
Agache, fig. 492: Agache-Bréart, 1975, 113.

RUE Bel., *Ambiani* (Somme, F) 50°16′N, 1°40°E
Fort, forming part of the shore defences (?), 'Les Fosses'. Sarcophagi, coins.
Agache, fig. 96, 620: Leduque, *Ambianie*, 161: Vasselle, 328: Agache-Bréart, 1975, 113–14: Agache, 1978, 24, 40, 247, 438, 452.

RUEIL Lug., *Parisii* (Hauts-de-Seine, F) 49°03′N, 1°53′E
RIGOIALENSIS/ROTOIALENSIS VICVS
Greg. Tur. *Hist Francorum* IX, 13; X, 28.
Gallo-Roman foundations.

Duval, 1961, 78, 231: Roblin, 1971, 8, 13, 26, 158, 259.

*RUMAISNIL Bel., *Ambiani* (Somme, F) 49°49′N, 2°08′E
Temple, 1st c, 'La Mare à chanvre'.
Agache, fig. 230, 439; 1972, 323: Agache-Bréart, 1975, 114: Agache, 1978, 267, 294, 295, 303, 337, 364, 365, 392, 396, 414.

RUMBERRY HILL (Langley) Brit., *Trinovantes* (Essex, GB) 51°59′N, 0°07′E
Tumulus, diameter at base 36.5 m; height 2.4 m. Bricks, terra sigillata, glass.
TEAS I (1858), 194: Fox, 1923, 196: *VCH Essex* III (1963), 152.

RVTTENV *cf* Ruan

RVTVPIAE *cf* Richborough

SABRINA FLVMEN *cf* river Severn

SACLAS Lug., *Carnutes* (Essonne, F) 48°22′N, 2°07′E
?SALIOCLITA
It. A. 368.1.
CIL I, p 670, n 8973.
Major settlement on the road between Paris and Orléans. Milestone now in Orléans museum. Settlement, end 1st c to end 4th c, vault, sculpture, 'Le creux de la Borne'.
Gallia XXX (1972), 303: Nouel, 5, 8.

SACLAY Lug., *Parisii* (Essonne, F) 48°44′N, 2°10′E
Major settlement on the road between Paris and Chartres; villa.
Duval, 1961, 232: Roblin, 1971, 143, 182, 192.

SAGII, SAII or ESVVII
Caes. *BG* II, 34; III, 7; V, 24, 53 (*Esuvios*).
Not. Gall. II, 6 (*civitas Saiorum*).
Not. Tiron. LXXXVII, 63 (*Saius*).
CIL XIII, no 630 (*Saiiae*).
Tribe of Lugdunensis (later Lugdunensis II) with capital at Sées (*qv*).

*SAINS-EN-AMIÉNOIS Bel., *Ambiani* (Somme, F) 49°49′N, 2°19′E
Gallo-Roman foundations; 'Camp de César'; altar with four divinities near the church.
Leduque, *Ambianie*, 149: *BSAP* VIII (1862–64), 261–65: Vasselle, 328: Agache-Bréart, 1975, 116: Agache, 1978, 438.

SAINT-ACHEUL-LÈS-AMIENS Bel., *Ambiani* (Somme, F) 49°54′N, 2°18′E
Cemetery, coins, small finds.
Leduque, *Ambianie*, 140, n 253, 263, 265–68: Agache-Bréart, 1975, 116: Agache, 1978, 21, 67, 68, 73, 202.

ST ALBANS Brit., *Catuvellauni* (Hertfordshire, GB) 51°45′N, 0°21′W
VERVLAMIVM
Coins of Tasciovanus (Mack, 152–92 *passim*).
Tac. *Ann.* XIV, 33 (*municipio Verulamio*).
Ptol. II,3,11 (Οὐρολάνιον).

87

It. A. 471.3 (*Verolamio*), 476.8 (*Verolami*), 479.8 (*Verolamo*).
Rav. Cos 106.50 (*Virolanium*).
Gildas, 10 (*Verolamiensis*).
Beda *Ecc. Hist.* I, 7 (*Verolamium*).
JRS XLVI (1956), 146–7.
PNRB 497–9.

Town, capital of the CIVITAS CATVVELLAVNORVM and possibly a *municipium*. Roads to London, Alchester, Towcester and Braughing. Belgic *oppidum* founded late 1st c BC. Roman fort, succeeded by town *c* AD 49; earth bank and ditch enclosing 48 ha, timber buildings. Destroyed by Boudica AD 60; rebuilt in Flavian period. Forum and basilica, theatre, temples, market, monumental arches. Earthen rampart (late 2nd c) enclosing *c* 90 ha; stone wall (mid 3rd c) enclosing *c* 80 ha. Place of the martydom of St Alban, early 3rd c; visited by St Germanus of Auxerre AD 429, at which time it was apparently still important. Some traces of occupation into 6th c. The mediaeval and modern town of St A. lies to the E of Roman *Verulamium*, separated from it by the river Ver. Museum.

Wheeler, R. E. M. and T. V., *Verulamium – a Belgic and Two Roman Cities*, Oxford, 1936: Frere, S. S. *Verulamium Excavations I*, Oxford, 1971, II, London, 1983: Rainey, 1973, 132–5: Wacher, 1975, 202–225; *RIB* 222–9.

SAINT-ANDRÉ-SUR-CAILLY Lug., *Veliocasses* (Seine-Maritime, F) 49°33′N, 1°13′E
Theatre, hamlet of Bouvent. Villa. Frankish tombs.
Cochet, *S.I.*, 198–201.

SAINT-AUBIN-CELLOVILLE *cf* Celloville

SAINT-AUBIN-D'ALGOT Lug., *Lexovii* (Calvados, F) 49°08′N, 0°05′E
Gallo-Roman foundations. Statuette of Mater Nutrix.
de Caumont, 1867, 413: *MSAN* XIX (1851), 37.

SAINT-AUBIN-DE-VIEIL-ÉVREUX *cf* Vieil-Évreux.

SAINT-AUBIN-SUR-GAILLON Lug., *Aulerci Eburovices* (Eure, F) 49°09′N, 1°20′E
Gallo-Roman composite sanctuary.
Coutil, *Archéol. g.r.*, Louviers, 259: Poulain, G., *Les fana ou temples g.r. de St-A* . . . Louviers, 1919: Grenier, *Manuel* IV.2 (1960), 742.

SAINT-AUBIN-SUR-MER Lug., *Viducasses* (Calvados, F) 49°20′N, 0°23′W
'le Camp Roman', 'le Castel', temple on which was superimposed a corridor villa with bathhouse, pits. Statuette of a mother goddess found broken in a pit; fragment of a terracotta tripod ornamented with a boat. Cemetery of about a dozen inhumations. Coins from 1st to 4th c. Terra sigillata signed VERONIVS.
de Caumont, 1846, 381: *BSAN* XLIX (1945), 398–99, 433–36, 445–56; L (1948), 234–39; LII (1949), 338: *MSAN* XLII-XLIV (1947–50), 83–97: *Gallia* VI (1948), 365–75: VII (1949), 121–22; IX (1952), 83–84.

SAINT-BIHY Lug., *Osismii* (Côtes du Nord, F) 48°23′N, 2°58′W
At 'Boissière' Gallo-Roman foundations, cement, tegulae.
MSCN I (1883–84), 266.

SAINT-BRANDAN Lug., *Coriosolites* (Côtes du Nord, F) 48°23′N, 2°52′W
'Le Rillan', probable *vicus* or minor settlement on the road between Carhaix and *Aletum* (St-Servan). Foundations, well with built edge, kilns, numerous coins of the early Empire; statue of Sucellos. 1st c.
MSCN I (1883–4), 267: Trevedy, 1892–95, 23: *Gallia* XXVIII (1970), 235–8.

SAINT-BRICE-SOUS-FORÊT Lug., *Parisii* (Val-d'Oise, F) 49°00′N, 2°22′E
Gallo-Roman foundations, 4th c.
Gallia XXVIII (1970), 250: Roblin, 1971, 175, 247.

SAINT-BRIEUC-DE-MAURON Lug., *Veneti* (Morbihan, F) 48°05′N, 2°22′W
'la Rosière', Gallo-Roman foundations, tegulae.
BSPM, 1972, 39.

SAINT-BROLADRE Lug., *Redones* (Ille-et-Vilaine, F) 48°35′N, 1°39′W
'le Clos Pottier', Gallo-Roman foundations, tegulae, coarse and fine pottery 2nd to 4th c.
Gallia XXXIII (1975), 341.

SAINT-CAST Lug., *Coriosolites* (Côtes du Nord, F) 48°38′N, 2°16′W
Gallo-Roman foundations, tegulae, on the road between la Chapelle de Saint-Cast and Vieuville.
BSCN XLVII (1909), 50.

SAINT-CHÉRON Lug., *Carnutes* (Essonne, F) 48°33′N, 2°07′E
Iron workings, furnace, slag.
Gallia IX (1951), 83.

SAINT-CLAIR-SUR-EPTE Lug., *Veliocasses* (Val-d'Oise, F) 49°12′N, 1°41′E
Gallo-Roman foundations, 'Beaujardin'; two important buildings with hypocausts, private baths, 2nd c.
Toussaint, 1951, 31–2: *Gallia* XIX (1961), 287: Roblin, 1978, 231.

SAINT-CLOUD-EN-DUNOIS Lug., *Carnutes* (Eure-et-Loir, F) 48°02′N, 1°28′E
Gallo-Roman foundations, 'Aux Murgers'; mosaics, tiles, coins, figurines.
Nouel, 16, 29.

SAINT-CÔME-DU-MONT Lug., *Unelli* (Manche, F) 49°20′N, 1°16′W
Gallo-Roman foundations, burials, tuyaux, tiles, statuettes of Venus; on the road between Carentan and Valognes: coin hoard from Augustus to Alexander Severus, 1st to mid-3rd c.
de Gerville, 1854, 182–3: *BSAIC* XXIV (1900), 145: *BSAN* LIII (1955–6), 197–263.

SAINT-COULOMB Lug., *Coriosolites* (Ille-et-Vilaine, F) 48°41′N, 1°54′W
'l'Anse de Duguesclin', Gallo-Roman foundations of a fish pond.
ASHAStM, 1966, 240–41: *Gallia* XXX (1972), 199–223: Langouet, 1973, 151.

ST DAVID'S HEAD Brit., *Demetae* (Dyfed, GB) 51°54′N, 5°16′W
OCTAPITARVM PROMONTORIVM.
Ptol. II,3,2 ('Οκταπίταρον ἄκρον).
PNRB 430.

SAINT-DENIS Lug., *Parisii* (Seine-Saint-Denis, F) 48°56′N, 2°22′E
?CATVLLIACVS (6th c).
Willeumier, 476 (milestone on the road to Rouen).
Cemetery, Gallo-Roman and Merovingian, under and around the basilica (five levels, the oldest mid-4th c). Dwellings, foundations of large structure (arch?), 3rd c.
CRAI, 1954, 391–400; 1958, 137–50: *Monuments Piot* XLIX (1957), 93–124: Salin, E., *Les tombes gallo-romaines et mérovingiennes de la basilique de St-D.,* Paris, 1958: Duval, 1961, 114, 223, 242, 281, 282: *BSNAF,* 1967, 230: Roblin, 1971, 119, 179, 206.

SAINT-DENOUAL Lug., *Coriosolites* (Côtes du Nord, F) 48°32′N, 2°24′W
'Les Clôtures', extensive Gallo-Roman foundations.
BSCN XLVII (1909), 51.

SAINT-DIDIER Lug., *Redones* (Ille-et-Vilaine, F) 48°06′N, 1°22′W
'les Suriaux', villa; building debris, local and imported pottery.
Gallia XXXIII (1975), 341.

SAINT-DIVY Lug., *Osismii* (Finistère, F) 48°27′N, 4°20′W
'Kerdalaun', Gallo-Roman foundations which might probably be those of a villa. Bronze statuettes; terra sigillata of the end of the 2nd and of the 3rd c.
du Chatellier, 1907, 129: *BSAF,* 1972, 99.

SAINT-FORGET Lug., *Parisii*? (Yvelines, F) 48°42′N, 2°00′E
Temple, Romano-Celtic, with smaller building, coins, mid-1st to end of 4th c; small finds, early Empire.
Paris et Île-de-France VIII (1956), 7–40: Grenier, *Manuel* IV.2 (1960), 718–21: Duval, 1961, 230: Roblin, 1971, 295.

SAINT-FRÉGANT Lug., *Osismii* (Finistère, F) 48°36′N, 4°23′W
'Kera dennec', a courtyard villa; baths, painted plaster, etc., on the road between Kérilien and l'Aberwrac'h. Metal and glass objects; coins from Augustus to Constantine, coin hoard of about 275. Abundant pottery including terra sigillata of the 2nd c and Argonne ware.
Ann Bret LXXVI (1969), 177–8; LXXVII (1970), 163, 225; LXXIX (1972), 167–214.

SAINT-GERMAIN-LÈS-CORBEIL *cf* Corbeil

SAINT-GILLES-VIEUX-MARCHÉ Lug., *Osismii* (Côtes du Nord, F) 48°15′N, 2°58′W
Gallo-Roman foundations; coins of early emperors.
MSCN I (1883–84), 530.

SAINT-GUINOX Lug., *Coriosolites* (Ille-et-Vilaine, F) 48°35′N, 1°53′W
Gallo-Roman foundations; bricks, potsherds, axes, vessels, bronze coins.
Chalmel, T., *Monographie de Saint-Guinox* (ad 1 f 1785), 100: Langouet, 1973, 151.

SAINT-HELEN Lug., *Coriosolites* (Côtes du Nord, F) 48°28′N, 1°58′W
At 'Croix du Fresne' Gallo-Roman foundations.
BSCN XLVII (1909), 21.

ST HILARY Brit., *Dumnonii* (Cornwall, GB) 50°08′N, 5°26′W
RIB 2233.
Milestone, of Constantine I as Caesar (AD 306–07).
Sedgley, 1975, 23–24.

SAINT-JEAN-AUX-BOIS Bel., *Suessiones* (Oise, F) 49°21′N, 2°55′E
Gallo-Roman foundations near a road.
Gallia XXV (1967), 199; XXVII (1969), 234.

SAINT-JEAN-DES-ÉCHELLES Lug., *Aulerci Cenomani* (Sarthe, F) 48°08′N, 0°43′E
Villa, 'Planchettes'; mosaics.
Bouton.

SAINT-JULIEN Lug., *Coriosolites* (Côtes du Nord, F) 48°27′N, 2°49′W
'Château de la Coste', tile kiln.
MSCN I (1883–84), 172.

SAINT-JUST-EN-CHAUSSÉE Bel., *Bellovaci* (Oise, F) 49°30′N, 2°26′E
SINOMO VICVS.
Vita S. Justi.
Villa. Place of the martyrdom of St Just.
DAG: Leduque, *Ambianie,* 150: Agache, 1978, 268: Roblin, 1978, 229–30, 319.

SAINT-LÉGER-DE-RÔTES Lug., *Lexovii* (Eure, F) 49°07′N, 0°39′E
Villa (and Frankish tombs).
Le Prévost, 1862, I, 48.

SAINT-LÉGER-DES-AUBÉES Lug., *Carnutes* (Eure-et-Loir, F) 48°25′N, 1°44′E
Gallo-Roman foundations, 'La Chasse'; ?pottery kiln.
Jalmain, air photograph: Nouel, 20.

SAINT-LEU-D'ESSERENT Bel., *Bellovaci* (Oise, F) 49°13′N, 2°25′E
Stone quarry and associated dwellings.
Gallia XVII (1959), 285.

SAINT-LÔ Lug., *Unelli* (Manche, F) 49°07′N, 1°05′W
Vicus or minor settlement; houses, aqueduct; bricks, tiles, a phial and tuyaux of terracotta, a weight of terra sigillata marked ANTOMAN, bull-horns. Coin hoard 1st, 2nd and 4th c.
Lambert, 1870, 105: *BSAIC* XXIV (1900), 147–8; supplement XXX (1908), 122–23: *BSAN* LIII (1955–56), 197–263: Bouhier, 1962, *v* St Lô.

SAINT-LORMEL Lug., *Coriosolites* (Côtes du Nord, F) 48°33′N, 2°14′W

At 'La Ville Orieu', Gallo-Roman foundations. At 'Château de l'Argentaye' a column with faint traces of a Roman inscription: discovered at Pluduno on the moor called 'la Milliaire', it marked the fourth league on the road between Corseul and Erquy.
BSCN XLVII (1909), 75.

SAINT-MADEN Lug., *Coriosolites* (Côtes du Nord, F) 48°20'N, 2°05'W
'La Rehaudais', Gallo-Roman foundations, tegulae, cement.
BSCN XLVII (1909), 14.

SAINT-MALO Lug., *Coriosolites* (Ille-et-Vilaine, F) 48°39'N, 2°01'W
ALETVM.
Not. Dig. Occ. XXXVII, 8 and 19 (*Praefectus militum Martensium, Aleto*).
Late fort. 'La Cité d'Alet', agglomeration of Gallo-Roman foundations. Very numerous remains beginning in the Gaulish epoch, at the greatest in the 4th c, ending early in the 5th c. Many coins and potsherds from Gaulish times to the late Empire.
Arche LXVIII (1974), 46–50: Johnston, 1977, 38–45: Johnson, 1979, 74–6, 81–2.

SAINT-MARCEL Lug., *Parisii* (Ville-de-Paris, F)
Christian *vicus*.
Duval, 1961, 232.

*SAINT-MARD-EN-CHAUSSÉE Bel., *Ambiani* (Somme, F) 49°41'N, 2°46'E
Gallo-Roman foundations, sarcophagi, tiles, coins, pottery.
Leduque, *Ambianie*, 143: Vasselle, 328: Agache, 1978, 41, 423.

SAINT-MARS-SOUS-BALLON Lug., *Aulerci Cenomani* (Sarthe, F) 48°10'N, 0°15'E
Gallo-Roman foundations, 'Épinais'; small finds, coins.
Bouton.

SAINT-MARTIN-DE-BRÉTHENCOURT Lug., *Carnutes* (Yvelines, F) 48°31'N, 1°55'E
Villa (large), outbuildings, mosaics, hamlet of Les Châtelliers, 'La Castille'.
Jalmain, air photograph.

SAINT-MARTIN-DES-CHAMPS Lug., *Osismii* (Finistère, F) 48°35'N, 3°50'W
'Bagatelle', cemetery; incinerations. Fibulae; coins from Augustus to Constantine.
du Chatellier, 1907, 85–86.

SAINT-MARTIN-OSMONVILLE Lug., *Caletes* (Seine-Maritime, F) 49°38'N, 1°18'E
Villa.
BCASI III (1873), 126–32 = *RA* XXVI (1873), 335–8: cf *Album Comm Ant SI* IV, 103.

SAINT-MAUR-DES-FOSSÉS Lug., *Parisii* (Val-de-Marne, F) 48°48'N, 2°30'E
Fort, late Empire; hypocaust, plaster. Municipal museum.
Duval, 1961, 227, 277, 284: Fort, E., *Catalogue du Musée de St-M*, 1934: *RÉA* XXII (1920), 107: Roblin, 1971, 216.

SAINT-MAUR-EN-CHAUSSÉE Bel., *Bellovaci* (Oise, F) 49°39'N, 1°55'E
Temple, large; double square set within a double circle.
Bull arch XXV–XXVI (1898), 92–96: Liebbe, M., *Rapport sur des fouilles pratiquées de l'Oise a St-M*.

SAINT-MAUR-LE-LOIR Lug., *Carnutes* (Eure-et-Loir, F) 48°09'N, 1°25'E
Gallo-Roman foundations, 'Edeville'; mosaics, terra sigillata, coins, small finds. Museums at Châteaudun and Bonneval.
Nouel, 16, 27.

SAINT-MAXIMIN Bel., *Bellovaci/Silvanectes* (Oise, F) 49°13'N, 2°27'E
?LITANOBRIGA.
It. A. 380.4.
Minor settlement, 'La Haute-Pommeraie', on the road between Beauvais and Senlis at the border between the *Bellovaci* and the *Silvanectes*. Weavers' and fullers' workshops, cellars (1st to 4th c), sanctuary, quarries. Villa, 'La Houy'. Oppidum with Gallo-Roman occupation, 'Canneville, Hironville'.
MAI XIV.2 (1942): *BSNAF*, 1954–5, 189; 1959, 93–4; 1971 (73), 51–60: Roblin, 1971, cf Pont-Ste-Maxence: *Bull Soc arch Creil*, Oct 1972, no 78: *Gallia* XVII (1959), 284; XIX (1961), 304; XXI (1963), 370; XXIII (1965), 323; XXV (1967), 197; XXVII (1969), 234; XXIX (1971), 228.

SAINT-MÉEN Lug., *Osismii* (Finistère, F) 48°34'N, 4°16'W
'Bouchéozen', Gallo-Roman foundations. One coin of the *Carnutes*; terra sigillata of 2nd and 3rd c.
BSAF, 1973, 74–76.

SAINT-MÉLOIR-DES-BOIS Lug., *Coriosolites* (Côtes du Nord, F) 48°27'N, 2°15'W
CIL XIII 9012.
Milestone, one of four columns opposite the church; dedicated by the *civitas* of the *Coriosolites* to the emperor M. Piavonius Victorinus (268–270).
BSCN XX (1862), 52–53; XXIX (1891), 188–89; XXXV (1897), 237; XLVII (1909), 78.

SAINT-MÉLOIR-DES-ONDES Lug., *Coriosolites* (Ille-et-Vilaine, F) 48°39'N, 1°54'W
'derrière la Haute Ville' Gallo-Roman foundations, tegulae, coarse and fine potsherds (Lezoux), fragment of amphora (Dressel 19–20); 2nd to 4th c.
Gallia XXXIII (1975), 343.

SAINT-MICHEL-DE-PLÉLAN Lug., *Coriosolites* (Côtes du Nord, F) 48°28'N, 2°13'W
'La Cheronnaix', important Gallo-Roman foundations, blocks of cement, tegulae.
MSCN I (1883–84), 472.

SAINT-NIC Lug., *Osismii* (Finistère, F) 48°12'N, 4°17'W
'Pentrez', salt manufactory on the Bay of Douarnenez near the road from Douarnenez to the Crozon peninsula.
du Chatellier, 1907, 258: *Gallia* XXX (1972), 217.

SAINT-NICOLAS-DU-BOSC Lug., *Lexovii* (Eure, F) 49°13'N, 0°52'E

Minor settlement. Temple? Tiles, bricks, pottery and coins found.
Le Prévost, 1862, III, 164–5.

ST OSYTH Brit., *Trinovantes* (Essex, GB) 51°48′N, 1°05′E
Salt workings, briquetage, pottery.
Nenquin, 1961, 89: *VCH Essex* III (1963), 176.

SAINT-OUEN Bel., *Ambiani* (Somme, F) 50°02′N, 2°07′E
Gallo-Roman foundations, 'Le Camp Fourillet'; tumulus, underground chambers.
Leduque, *Ambianie*, 127: Agache-Bréart, 1975, 116: Agache, 1978, 116, 345.

SAINT-OUEN-DE-THOUBERVILLE Lug., *Lexovii* (Eure, F) 49°21′N, 0°53′E
Temple.
Coutil I (*Pont-Audemer*), 102; II (*Louviers*), 123: Vesly, 20: Deglatigny, L., *Notes sur le temple gallo-romain de St-O.-de-T.*, Rouen, 1922: Grenier, *Manuel* III.1 (1958), 464–7: Duval, 1961, 230.

SAINT-PAUL-DE-COURTONNE Lug., *Lexovii* (Calvados, F) 49°05′N, 0°28′E
Small villa; hypocaust with semicircle paved with terracotta; conduits in terracotta; marble capitals, painted plaster.
BSAN II (1862), 473–75.

SAINT-PIERRE-DE-SÉMILLY Lug., *Unelli* or *Baiocasses* (Manche, F) 49°07′N, 1°00′W
Gallo-Roman foundations and coin hoard of the early Empire.
BSAN LIII (1955–56), 197–263.

SAINT-PIERRE-LA-GARENNE Lug., *Aulerci Eburovices* (Eure, F) 49°09′N, 1°24′E
Villa. Romano-Frankish cemetery near the ancient Chappelle N.-D. de la G.
Coutil, II (*Louviers*), 273: Grenier, *Manuel* II.2 (1934), 870.

SAINT-PIERRE-LA-VIEILLE Lug., *Viducasses* (Calvados, F) 48°55′N, 0°34′W
Gallo-Roman foundations; road.
BSAN VI (1872), 209: *Ann 5 Norm*, 1957, 59–67.

SAINT-PIERRE-LÈS-NEMOURS Lug., *Senones* (Seine-et-Marne, F) 48°16′N, 2°41′E
Gallo-Roman foundations, bronze vessels.
Nouel, 21, 33, 36.

SAINT-POTAN Lug., *Coriosolites* (Côtes du Nord, F) 48°33′N, 2°17′W
CIL XIII, 8992.
At 'Saint-Maudez', on the road to Saint-Lormel and not that to Ruca as indicated by Harmois, a milestone: TRIB(unicia) (p)OT(estate) SEPTIMA CO(n)SVL. Destroyed in 1852.
At 'La Haugue-Morais' a bronze statuette bearing a sacrificial knife; an aureus of Trajan, 97–117.
BSCN V (1867), appendix 132; XXXV (1897), 251; XLVII (1909), 52.

SAINT-QUAY-PORTRIEUX Lug., *Osismii* (Côtes du Nord, F) 48°39′N, 2°50′W
Gallo-Roman foundations and debris; one bronze coin each of Nero and Gallienus.
MSCN I (1883–84), 221.

SAINT-QUENTIN-LA-ROCHE Lug., *Viducasses* (Calvados, F) 48°58′N, 0°14′W
Oppidum, with temple and villa on the road between Avranches and Jort. The temple near the church. Fragment of a bronze tripod terminated by a lioness; statuette of Venus Anadyomene. Scattered coins from Gaulish to Honorius; a hoard of coins between 268 and 275. Terra sigillata including MARTI. DONTIO FIICI. VF MAT. ATEPOM. . . .LI M.
de Caumont, 1850, 321: *BSAN* XXIX (1913), 229; LII (1955), 109: *BSNÉP* XXXIII (1942), 52–59.

SAINT-RÉMY-DES-MONTS Lug., *Aulerci Cenomani* (Sarthe, F) 48°18′N, 0°24′E
Villa, 'Les Terres Noires'; funerary pits, coins.
Bouton.

SAINT-RIQUIER Bel., *Ambiani* (Somme, F) 50°08′N, 1°57′E
CENTVLA (7th c.).
Gallo-Roman remains, burials, small finds.
Agache, fig. 118, 177: Leduque, *Ambianie*, 172, n. 839–42: Agache-Bréart, 1975, 117: Agache, 1978, 145, 204, 231, 257, 269, 315.

SAINT-RIQUIER-EN-RIVIÈRE Lug., *Caletes* (Seine-Maritime, F) 49°53′N, 1°34′E
Glassworks in the forest of Eu.
Leduque, *Ambianie*, n. 149.

SAINT-SAËNS Lug., *Caletes* (Seine-Maritime, F) 49°40′N, 1°17′E
Temple du Tertre.
Vesly, 16.

ST STEPHEN Brit., *Catuvellauni* (Hertfordshire, GB) 51°43′N, 0°22′W
Tile kiln; 2nd c. ('Blackboy Pits').
TStAHAAS ns IV (1932), 212–14.

SAINT-SULIAC Lug., *Coriosolites* (Ille-et-Vilaine, F) 48°34′N, 1°58′W
At 'Anse de Vigneux' Gallo-Roman foundations with hypocaust, tegulae, local pottery. Also others, including a hypocaust and 3rd c local pottery.
Langouet, 1973, 151: *Gallia* XXXIII (1975), 343.

SAINT-VAAST-LÈS-MELLO Bel., *Bellovaci* (Oise, F) 49°16′N, 2°23′E
Stone quarry, villa.
Gallia VII (1949), 114; XVII (1959), 285.

SAINT-VALÉRY-SUR-SOMME Bel., *Ambiani* (Somme, F) 50°11′N, 1°38′E
?LOCVS QVARTENSIS SIVE HORNENSIS
Not. Dig. Occ. XXXVIII, 8 (*praefectus classis Sambricae in loco Quartensi sive Hornensi*).
Gallo-Roman foundations, columns, ?temple, tiles, pottery, small finds. In the mud of the former port, a Neronian ingot of lead from the mines of Somerset.
The location of *Locus Q. sive H.* is disputed: le Crotoy, Étaples, Hargnies, Hornoy, Cap Hornu (with a Gallic oppidum), Marchiennes, Pont-sur-Sambre (Quartes), Quaregnan-Hornu?

MSÉA VI (1849), 176: *RA* XIII.1 (1921), 36: Jullian, *HG* VIII (1926), 107: Grenier, *Manuel* I (1931), 390; II.1 (1934), 251: Agache, fig. 2: Vasselle, 329: Leduque, *Ambianie*, 165, n 741, 745: Agache-Bréart, 1975, 117: Johnson, 1979, 92–3.

SAINT-VIGOR-LE-GRAND Lug., *Baiocasses* (Calvados, F) 49°17′N, 0°41′W
'Colline St. Floxel', temple – a pagan sanctuary beneath the abbey. Cinerary urns and sarcophagi, a child's sarcophagus hollowed out of a Constantinian milestone. Statuettes in terracotta; coins from 2nd to 4th c.
de Caumont, 1857, 449: *Gallia* XXIV (1966), 259–60.

SAINT-VRAIN Lug., *Parisii* (Essonne, F) 48°32′N, 2°20′E
Villa; late-Empire pottery, objects of art, jewels, coins 2nd to 4th c.
Duval, 1961, 229: Roblin, 1971, 175, 305.

SAINTE-CATHERINE-LÈS-ARRAS Bel., *Atrebates* (Pas-de-Calais, F) 50°18′N, 2°46′E
Kiln.
Dérolez, 516.

SAINTE-CROIX-HAGUE Lug., *Unelli* (Manche, F) 49°38′N, 1°46′W
Small villa at 'rue d'Ozouville'; Gaulish and Gallo-Roman pottery kilns; tiles, querns, jet figure 5 cm high, bronze spoon, potsherds.
BSAIC XXIV (1900), 51–5: *BSNÉP* XIII (1905), 184: Bouhier, 1962, *v* Ste Croix-Hague.

SAINTE-GENEVIÈVE-DES-BOIS Lug., *Parisii* (Essonne, F) 48°38′N, 2°20′E
Gallo-Roman foundations, late Empire, statuette of Mercury, tiles.
Duval, 1961, 229: Roblin, 1971, 183, 297.

SAINTE-MARGUERITE-SUR-MER Lug., *Caletes* (Seine-Maritime, F) 49°55′N, 0°57′E
Temple, villa.
Bull Mon IX (1843), 92–7: Cochet, *S.I.*, 242: *BCASI* IV (1876–8), 68–74: *BSAN* VIII (1878): Vesly, 12: Grenier, *Manuel* II.2 (1934), 868.

SAINTE-MARIE-DU-MONT Lug., *Unelli* (Manche, F) 49°23′N, 1°13′W
Small villa, 'la Sagerie', at a cross-roads; tiles, bricks. Coins, among them a gold coin of Constantine. 2nd and 3rd c.
de Gerville, 1854, 185: *BSAIC* XXIV (1900), 91–2; supplement XXX (1908), 96.

***SAINTE-SEGRÉE** Bel., *Ambiani* (Somme, F) 49°45′N, 1°55′E
Gallo-Roman foundations, sarcophagi, coins, pottery, tiles, querns.
Leduque, *Ambianie*, 157, n. 568–70.

***SALEUX** Bel., *Ambiani* (Somme, F) 49°52′N, 2°15′E
Cemetery, coins, pottery, small finds.
Agache, fig. 337: Leduque, *Ambianie*, 156, n. 548–49: Agache-Bréart, 1975, 117: Agache, 1978, 38, 165.

SALIOCLITA *cf* ?Saclas.

SAMARA FLVMEN *cf* river Somme.

SAMAROBRIVA *cf* Amiens.

SAMER Bel., *Morini* (Pas-de-Calais, F) 50°38′N, 1°45′E
Major settlement, temple, sacred fountain.
Leduque, *Boulonnais*, 99.

SANDY LANE Brit., *Belgae* (Wiltshire, GB) 51°24′N, 2°03′W
VERLVCIO
It. A. 486.4.
PNRB 494.
Minor settlement, on the road between Silchester and Bath. Many small finds, tesserae, coin of Victorinus.
VCH Wiltshire I (1957), 53–54.

SANGATTE Bel., *Morini* (Pas-de-Calais, F) 50°56′N, 1°45′E
Port (settlement submerged); Gallo-Roman foundations, coins. Possibly the *superior portus* of Caesar, B.G. IV, 28.
Leduque, *Boulonnais*, 112, n. 14; *Morinie*, 71. *Procs. Brit. Academy* LXIII (1977), 150–2.

SANGHEN Bel., *Morini* (Pas-de-Calais, F) 50°47′N, 1°54′E
Cemetery, pottery, small finds.
Ringot, 176.

SARS (Le) Bel., *Atrebates* (Pas-de-Calais, F) 50°04′N, 2°47′E
Minor settlement.
Dérolez, 521: Agache-Bréart, 1975, 141: Agache, 1978, 246.

SAUMERAY Lug., *Carnutes* (Eure-et-Loir, F) 48°15′N, 1°19′E
Gallo-Roman foundations, 'Bû'; pits.
Gallia XXI (1963), 396.

SAUNDERTON Brit., *Catuvellauni* (Buckinghamshire, GB) 51°43′N, 0°51′W
Villa; hypocausts. Occupied 2nd c.
Records of Bucks XIII (1939), 398–426: XVIII (1969), 261–76: *Arch J* CXXIV (1967), 129–159, *passim*.

SAUNDERTON LEE Brit., *Catuvellauni* (Buckinghamshire, GB) 51°41′N, 0°51′W
Villa; basilical. Overlies Iron Age settlement.
JRS LV (1965), 88: LXIII (1973), 246: *Arch J* CXXIV (1967), 129–159, *passim*.

SAUSSEMESNIL Lug., *Unelli* (Manche, F) 49°34′N, 1°27′W
Gallo-Roman foundations; manufacture of coarse pottery; coins of 1st and 2nd c.
De Gerville, 1854, 200: Renault, 1867, 53.

SAVERNAKE FOREST Brit., *Atrebates* (Wiltshire, GB) 51°23′N, 1°41′W
Pottery kilns, late 1st and early 2nd c.
WANHM LVIII (1961–63), 142–55.

SCEAUX-DU-GÂTINAIS Lug., *Senones* (Loiret, F) 48°06′N, 2°36′E
?AQVAE SEGESTAE
Tab. Peut. (*Aquis Segeste*).

Major settlement (spa) on the road between Sens and Orléans. Baths, aqueduct, theatre, temple, mosaics (late 1st to 3rd c), statuette, coins.

Desjardins, *Tab. Peut.*, 26: Grenier, *Manuel* III.2 (1958), 874–6: *Gallia* XXIV (1966), 241; XXVI (1968), 325; XXVIII (1970), 260: Nouel, 18: Jalmain, air photograph of important structures west of the excavations of Préau: *BSNAF*, 1972 (74), 192–3.

SCEAUX-SUR-HUISNE Lug., *Aulerci Cenomani* (Sarthe, F) 48°06′N, 0°35′E

Baths, 'La Cour', with vestibule corridor, hypocausts, *praefurnium, frigidarium, tepidarium, caldarium, sudatio*, coins, small finds. At 'Roches', brick walls, three structures, hypocaust, traces of burning, coins of the sons of Constantine.

Bull Mon, 1868, 292: Bouton: Charles, L. and R., *Hist. de la Ferté-Bernard*, Le Mans, 1877, 81 sq: Ledru, 1911, 300–3: Pesche, *Dictionnaire* VI, 90.

SCILLY ISLES Brit., *Dumnonii* (Cornwall, GB) 49°55′N, 6°20′W
?SILINA INSVLA
?Plin. *NH* IV, 103 (*Silumnus* or *Silimnus*).
?Solinus 22.7 (*Siluram*).
Sulpicius Severus II, 51, 3–4 (*Sylina, Sylinancim*). *PNRB* 457–9.
Roman pottery, coins and many brooches from native shrine (?) at Nor'nour.
Arch. J. CXXIV (1967), 1–64.

SEAFORD Brit., *Regni* (East Sussex, GB) 50°46′N, 0°06′E
Salt working, briquetage, pottery.
Nenquin, 1961, 90.

SEA MILLS Brit., *Belgae* (Avon, GB) 51°29′N, 2°39′W
ABONA
It. A. 486.1 (*Abone*).
Rav. Cos. 106.22 (*Punctuobice* = ?*Portu Abone*).
PNRB 240.
Large settlement and port; ferry across the river Severn. Occupied Claudian period to *c* AD 120, then again from late 3rd to early 5th c. Walls, floors, pavements, many small finds, 4th-c coins.
TBGAS LXVI (1945), 258–95; LXVIII (1949), 184–8.

SEATON Brit., *Durotriges* (Devonshire, GB) 50°43′N, 3°05′W
Villa at Honeyditches; timber buildings, small bath-house.
Fox, A., *South West England*, Newton Abbot, 1973, 172: *Britannia* VIII (1977), 107–148.

SEAVINGTON ST MARY Brit., *Durotriges* (Somerset, GB) 50°55′N, 2°51′W
Villa; mosaics, hypocausts; coin of Tetricus (AD 270).
VCH Somerset I (1906), 332: *SDNQ* XXV (1949), 239–40.

SÉES Lug., *Esuvii* or *Sagii* (Orne, F) 48°36′N, 0°10′E
?NVDIONNVM or NVTIONNVM, later SAGIIS
Tab. Peut. (*cf* Jublains).

Town on the road between Le Mans and Vieux, capital of the CIVITAS SAGIORVM under the late Empire. Polygonal temple, sculpture.
Hommey, Abbé L., *Histoire . . . du diocèse de Séez* . . ., Alençon, 1898–1900: *Ann Norm* XVI (1966), 384–7: *Gallia* XXVI (1968), 367.

*SEEVIAE (STEVIAE, SELVIAE), SEFVLAE
Tab. Peut. SETVCIS.
Milestone at Tongres (*CIL* XIII, 2, 2, p. 711, 9158) SEFVLAE.
Ambianian settlement on the road between Amiens and Soissons, location uncertain: Hangest-Quesnel – or towards Fresnoy-en-Chaussée – (Somme) or near Mézières or Demuin (Miller).
An air photograph by Agache shows very numerous small habitations at the deserted village of St-Mard, between Fresnoy and Beaucourt.
Agache-Bréart, 32; Agache, 1978, 41, 423.

SEGLIEN Lug., *Veneti* (Morbihan, F) 48°07′N, 3°09′W
CIL XIII 9009, 9010 (1 and 2).
At 'St Germain', 'St Jean' and 'St Zenon' on the road between Carhaix and Rennes, mutilated milestones. At 'Loucouviern' Gallo-Roman foundations forming a long rectangle and enclosing debris of tegulae, lustrous black pottery and many querns. At 'Guergomel', 'Maneguegan', Gallo-Roman foundations, tegulae.
BSPM, 1908, 63–68; 1912, 71; 1936, 64; 1945, 112: *Ann Bret* LXII (1955), 332.

SEINE, river (F)
SEQVANA FLVMEN
Caes. *BG* I, 1; VII, 57, 58.
Str. IV,1,14 (C. 189) (Σηκοάνας).
Mela III, 20.
Plin. *NH* IV, 105, 109.
Ptol. II,8,2, 3, 5, 7, 8, 9, 10; 9, 1, 4, 6.
Cass. Dio XL,38,4.
Pan. Const. Caes. XIV.
Amm. Marc. XV,11,3.
Marcian. II, 24, 25, 26, 27, 30.
Vib. Seq.
Sid. Apoll. *Carmina* V, 208.
Ven. Fort.
Greg. Tur.
CIL XIII.1, p 437 (notes); 2858–64 (invocations to *Dea Sequana*); XIII.2, 5595 (*Sequ(anam) vers(us)*).
Duval, 1961, index, *sv*.

SENA INSVLA *cf* Île de Sein

SENANTES Lug., (Eure-et-Loir, F) 48°40′N, 1°34′E
Temple of square plan, villa, other structures, hypogeum, of Flavian-Antonine period, at the boundary between the *Carnutes* and the *Eburovices*.
Gallia XXVI (1968), 324.

SENLIS Bel., *Silvanectes* (Oise, F) 49°12′N, 2°35′E
RATOMAGVS (?) or AVGVSTOMAGVS SILVANECTVM, later SILVANECTAS
?Ptol. II,9,6 ('Ρατόμαγος).

It. A. 380.5 (*Augustomago*).
Tab. Peut. (*Aug. Magus*).
CIL XIII.2, p 543.
Cf RATOMAGVS and SILVANECTES.
Town, capital of the CIVITAS SILVANECTVM. Roads to Beauvais, Paris, Sens and Soissons. Rectangular street-plan, with *cardines* and *decumani*, covering c 45 ha; forum, baths, temples, market, amphitheatre, 3rd-c walls enclosing c 6.5 ha. Musée du Haubergier.
RÉA V (1903), 35; LXVII (1965), 368–91: Blanchet, *Enceintes*, 112–6: Grenier, *Manuel* I (1931), 414, 532; III.1 (1958), 246–9; III.2 (1958), 886–90: *Mém Soc Senlis*, 1940–45, 1952: *BSNAF*, 1942, 196–212; 1964, 72–4: Louat, F., *Histoire de la ville de Senlis*, 2nd ed., S., 1944: *CR Soc Senlis*, 1946–7, 9–12: *Gallia* V (1947), 439; VII (1949), 114; IX (1951), 83; XII (1954), 146; XV (1957), 165; XIX (1961), 301; XXI (1963), 369; XXIX (1971), 228: *DAC* XV (1950), col. 1200–1240: *CRAI*, 1959 (1960), 450–7: *Actes du 90e Congrès nationale des Sociétés savantes*, Nice, 1965 (1966), 221: *Britannia* IV (1973), 210–23: Roblin, 1978, 209–13.

SENONES
Caes. *BG* II, 2; V, 54, 56; VI, 2, 3, 5, 44; VII, 4, 11, 34, 56, 58, 75; VIII, 30.
Str. IV,3,5 (C. 194) (Σένονες).
Plin. *NH* IV, 107.
Ptol. II,8,9.
Conc. Agrip. a. 346.
Eutrop. X,12,2.
Amm. Marc. XV,11,11; XVI,3,3.
Not. Gall. IV, 2 (*metropolis civitas Senonum*).
Iul. Hon. XIX.
Oros. VII, 29.
Ven. Fort. *Vita S Martini* IV, 173.
Greg. Tur. *HF* VIII, 31; X, 11.
CIL VII, 191 = RIB 262; XIII, 921, 1676, 1684; notes, p 443; 2675, 2924, 2942, 2949, 3067; XIII.2, 6084.
Large tribe of Celtic Gaul, included in Lugdunensis (later Lugdunensis Senonia). Capital: AGEDINCVM, later known as SENONAS (Sens). Other settlements: AVTESSIODVRVM (Auxerre), INTERAMNVM (Entrains), VICVS MASAVENSIVM (Mesves), METIOSEDVM (Melun) and VELLAVNODVNVM, of which only the last two appear on this map. Another branch of the tribe entered Italy c 400 BC and was famous for its sack of Rome in 390.
Bull Soc sc Yonne, 1935, 239; 1936, 258; 1937, 249; 1938, 108 ff: *RAE* XXIV (Oct-Dec 1955), 334.

SEQVANA FLVMEN cf river Seine
SERMAISES-EN-BEAUCE Lug., *Carnutes* (Loiret, F) 48°18'N, 2°12'E
Four villas, one south of the town at 'la Haie d'Argeville'; cemetery, coin hoard.
Jalmain, air photograph: Nouel, 5, 9, 27, 35; *Bull Soc arch Orl* XX, no. 26.
SERQUIGNY Lug., *Lexovii* (Eure, F) 49°06'N, 0°43'E

Villa.
de Caumont, 1838, 162: Le Prévost, 1862, I, 45–6; III, 239.
SETVCIS cf *Seeviae*, ?Fresnoy-en-Chaussée.
SEVERN, river Brit., *Dobunni* at mouth (Gloucestershire, GB)
SABRINA FLVMEN.
Tac. *Ann.* XII,31.
Ptol. II,3,2 (Σαβρίνα εἴσχυσις).
Gildas, 3.
Beda *Ecc. Hist.*, V,23.
Hist. Brit. II.9, II.49, VII.68, VII.69, VII.72.
PNRB 450.
SHAKENOAK Brit., *Dobunni* (Oxfordshire, GB) 51°49'N, 1°27'W
Villa; site A occupied AD 100–430; aisled barn later villa; baths. Site B occupied 1st to 5th c, a corridor villa. Numerous finds. Saxon occupation lasting to 8th c.
Brodribb, A. C. C., Hands, A. R. and Walker, D. R., *Excavations at Shakenoak*, four volumes, Oxford, 1968–1973.
SHAPWICK Brit., *Durotriges* (Dorset, GB) 50°48'N, 2°04'W
Fort on the road from Badbury to Dorchester (Dorset) near the crossing of the river Stour. Area c 2.4 ha. Probably connected with Vespasian's campaign against the *Durotriges*, AD 43 (cf Lake Farm, Hod Hill).
Britannia VII (1976), 280–83: Webster, 1980, 145.
SHEDFIELD cf Hallcourt Wood
SHEPTON MALLET Brit., *Belgae* (Somerset, GB) 51°11'N, 2°33'W
Pottery, five kilns.
VCH Somerset I (1906), 317–19.
SHIPSTAL Brit., *Durotriges* (Dorset, GB) 50°41'N, 2°02'W
Salt working, 1st and 2nd c.
RCHM Dorset II part 3 (1970), 593.
SHOEBURY Brit., *Trinovantes* (Essex, GB) 51°32'N, 0°48'E
Four pottery kilns.
VCH Essex III (1963), 178–9.
SIBLE HEDINGHAM Brit., *Trinovantes* (Essex, GB) 51°58'N, 0°35'E
Two pottery kilns, one 4th c.
VCH Essex III (1963), 145.
SIDLESHAM Brit., *Regni* (West Sussex, GB) 50°46'N, 0°47'W
Villa; occupied 2nd to 4th c. Baths, both wet- and dry-heat types, cf Chedworth.
SxAC CXI (1973), 1–19.
SILCHESTER Brit., *Atrebates* (Hampshire, GB) 51°21'N, 1°05'W
CALLEVA ATREBATVM.
Belgic coins (*Callev* or *Rex Calle*: *Arch* X (1944), 7–8: Mack, nos. 107, 108).
Ptol. II,3,12 (Καληοῦα).

It. A. 478.3 (*Galleva Atrebatum*), 484.10, 485.7, 485.8, 486.7, 486.8 (*Calleva*).
Rav. Cos. 106.32 (*Caleba Arbatium*).
RIB 67–87, 2221.
PNRB 291.

Town, capital of the CIVITAS ATREBATVM. Roads to London, Chichester, Winchester, Dorchester (Dorset), Bath, Cirencester and Alchester. The pre-Roman capital of the *Atrebates*, with a mint, occupied from early 1st c AD, abandoned early 5th c. Inner earthen rampart (c AD 40–50) encloses 32.5 ha; outer earthen rampart (Flavian?) encloses 95 ha; a stone wall, much of which remains, was built in 3rd c. to face a third (late 2nd c.) earthwork which enclosed c 40 ha. Forum, basilica, amphitheatre, baths, temples, many shops and houses, probable Christian church. There may have been a port for Silchester on the river Thames at Reading.

Rainey, 1973, 136–40: Boon, C. G., *Silchester*, Newton Abbot, 1974, *passim*: Wacher, 1975, 255–77. *Arch.* CII (1969), 1–82; CV (1975), 277–302; *Britannia* IX (1978), 464–5; X (1979), 331; XI (1980), 394–6.

SILIA INSVLA *cf* ?Guernsey

SILINA INSVLA *cf* ?Scilly Isles

SILVRES
Plin. *NH* IV, 103.
Tac. *Agric* 11,2; 17,3: *Ann* XII, 32, 33, 38, 39, 40; XIV, 29.
Ptol. II,3,12 (Σίλυρες).
It. A. 485.9 (*Venta Silurum*).
Rav. Cos. 106.22 (*Ventaslurum*).
Jordanes II, 13.
CIL II, 5923 (*Silur*).
RIB 311 (*res publ*(*ica*) *civit*(*atis*) *Silurum*).
PNRB 459–60.

British tribe occupying Gwent and the Glamorgans. Consistently hostile to Rome, mid 1st c AD; finally conquered by Frontinus, c 75; afterwards established as a *civitas* with capital at VENTA SILVRVM (Caerwent, *qv*).

Rivet, 1964, 160: Nash-Williams, 1969: Wacher, 1975, 375–6: Frere, 1978, 120–22 etc.: Salway, 1981, 137–8.

SILVANECTES
Plin. *NH* IV, 106 (*Ulmanectes/Ulmancetes/Ulmanetes, liberi*).
Ptol. II,9,6 (Οὐβάνεκτοι/Σουβάνεκτοι/Σουμάνεκτοι).
Not. Dig. Occ. XLII, 42 (*Silvanectas*).
Not. Gall. VI, 9 (*civitas Silvanectum*).
Conc. Aurel. a. 511 (*episcopus de Silvanectis*).
Greg. Tur. *HF* VI, 8(14), 33(46); IX, 20 (*Silvanectensis*).
CIL XIII.1, p 543.
CRAI, 1959, 450 (*civitas Sulbanectium*). Dedication of city temple: *cf* Bath (AQVAE SVLIS, with temple of *Sulis Minerva*).

Gallic tribe not mentioned by Caesar, perhaps in his time a dependency of the *Meldi*. Established as a *civitas* of Gallia Belgica, probably by Augustus. Capital: AVGVSTOMAGVS, later known as SILVANECTAS (Senlis, *qv*).

Roblin, M., Les limites de la *civitas* des S., *JS*, 1963, 65–85; 1965, 543–63.

SIZUN Lug., *Osismii* (Finistère, F) 48°24′N, 4°05′W
'Kergréac'h', small villa, little farm building in the form of a rectangle. Local and imported pottery of the 2nd c.
Gallia XIII (1955), 156–57.

SLODEN *cf* Crock Hill

SNODLAND Brit., *Cantiaci* (Kent, GB) 51°19′N, 0°26′E
Villa, corridor; bath, wall-paintings, mosaics.
Arch Cant LXXXII (1967), 192–217.

SOMBRIN Bel., *Atrebates* (Pas-de-Calais, F) 50°14′N, 2°30′E
Fort.
Dérolez, 521.

SOMME, river Bel., *Ambiani* at mouth (Somme, F)
SAMARA FLVMEN (later SOMENA, SVMENA, SVMINA, SVMNA, SVMMAMA).
Ptol. II,9,1 (Φρούδιος ποταμοῦ ἐκβολαί).
Not. Dig. Occ. XXXVIII, 8 (*Classis Sambrica*).
CIL XIII.1, p 561, col II.
(Amiens = SAMAROBRIVA = 'Sommebridge').

SORVIODVNVM *cf* Old Sarum

SOTTEVAST Lug., *Unelli* (Manche, F) 49°31′N, 1°35′W
Small villa at 'le Clod de l'Hôtel'; conduits belonging to an aqueduct. A bronze vessel, scattered coins and a hoard of 3000 silver coins from Vespasian to Postumus; one gold coin of Galla Placidia.
MSAN V (1829–30), 326–30: de Gerville, 1854, 201: Blanchet, 1900, 206 n 426: *BSAIC* XXIV (1900), 96–97: *BSAN* LIII (1955–56), 197–263: Bouhier, 1962, *v* Sottevast.

SOURS Lug., *Carnutes* (Eure-et-Loir, F) 48°25′N, 1°36′E
Aqueduct, 'Lièvreville'; Gallo-Roman foundations.
Jalmain, air photograph: Nouel, 26.

SOUTH FORELAND Brit., *Cantiaci* (Kent, GB) 51°08′N, 1°22′E
CANTIVM PROMONTORIVM.
Diod. Sic. V,21,3 (ἀκρωτήριον ... Κάντιον).
Str. I, 4, 3 (C. 63) etc.(Κάντιον).
Ptol. II,3,3–4 (Κάντιον ἄκρον).
PNRB 300.

SOUTH OCKENDON Brit., *Trinovantes* (Essex, GB) 51°30′N, 0°18′E
Tumulus; diameter at base 35.7 m, surrounded by a ditch. Pottery, querns, 3rd c.
TEAS XXV (1958), 271–72.

SOUTHWICK Brit., *Regni* (West Sussex, GB) 50°50′N, 0°14′W
Villa; courtyard. Occupied c 100 (possibly earlier) to 350. Baths, mosaics, painted plaster, coins

Hadrian to Constantine.
SxAC LXXIII (1932), 13–32: Cunliffe, 1973, 77–8.

SOUZY-LA-BRICHE Lug., *Carnutes* (Essonne, F) 48°32′N, 2°09′E
Large villa with other structures, mosaics. Museum, Étampes.
Jalmain, air photograph.

SPINIS Brit., *Atrebates* (Berkshire, GB) 51°24′N, 1°21′W
It. A. 485.6; 486.6.
PNRB 462.
Ill-defined minor settlement located by mention in *It. A.*, apparently centred near Woodspeen Farm.
Peake, H., *Archaeology of Berkshire*, London, 1931, 229: *Britannia* I (1970), 58, 79.

SPOONLEY WOOD Brit., *Dobunni* (Gloucestershire, GB) 51°56′N, 1°56′W
Villa, courtyard. Baths, many mosaics, fragments of columns, pottery, coins. Shrine (Cole's Hill), uninscribed altar, part of stone effigy of an eagle in local limestone, brooches, bracelets, rings, coins of 3rd and 4th c; inhumations.
Arch LII (1889–90), 651–68: *TBGAS* LXXI (1953), 162–66: *JRS* XLIX (1959), 127: Rainey, 1973, 144–45: McWhirr, 1981, 94–6; RCHM, *Gloucestershire Cotswolds*, 1976, 113–4.

SPRINGHEAD Brit., *Cantiaci* (Kent, GB) 51°25′N, 0°18′E
?VAGNIACIS
It. A. 472.2.
PNRB 485.
Town on the road between London and Rochester, centre of cult of springs. Seven Romano-Celtic temples, remains of many other buildings; pottery kiln.
Arch Cant LXXX (1965), 107–17: Ross, 1967, 20–33: *JRS* LIX (1969), 232: *Britannia* I (1970), 44, 79: Rodwell-Rowley, 1975, 39.

STAINES Brit., *Catuvellauni* (Surrey, GB) 51°26′N, 0°31′W
PONTES
It. A. 478.4 (*Pontibus*).
PNRB 441.
Minor settlement at the crossing of the river Thames on the road between London and Silchester. Timber buildings, pottery, coins, many small finds.
VCH Middlesex I (1969), 73: *LA* I (1970), 161–2: *TLMAS* XXVII (1976), 71–134.

STAFFORD COMMON Brit., *Silures* (West Glamorgan, GB) 51°39′N, 4°02′W
Practice camp.
JRS LI (1961), 119–35; RCAHM (Wales), *Glamorgan* I, ii, 1976, 101.

STAMPAE *cf* Étampes

STANCOMBE PARK Brit., *Dobunni* (Gloucestershire, GB) 51°40′N, 2°22′W
Villa; foundations, mosaic.
JBAA II (1847), 349; RCHM, *Gloucestershire Cotswolds*, 1976, 111.

STANDEN Brit., *Regni* (West Sussex, GB) 51°06′N, 0°01′W
Iron working; pottery 2nd c.
SxNQ VII (1938–39), 153–54.

STANTON PARK Brit., *Dobunni* (Wiltshire, GB) 51°31′N, 2°09′W
Villa; hypocausts, *tesserae*, tiles, painted plaster.
VCH Wiltshire I (1957), 107.

STANWAY Brit., *Trinovantes* (Essex, GB) 51°51′N, 0°50′E
Fort, area *c* 2 ha; earthen rampart, timber buildings. Probably built immediately after the conquest of *Camulodunum*, AD 43. *cf* Gosbeck's Farm.
JRS LXVII (1977), 126–8: *Britannia* VIII (1977), 185–7; Webster, 1980, 129–31.

STEVENAGE Brit., *Catuvellauni* (Hertfordshire, GB) 51°54′N, 0°12′W
Six barrows; said once to have had ditches with external banks. Mutilated.
TEHAS III (1905–07), 178–85: Dunning, G. C. and Jessup, R. F., *Roman Barrows*, London, 1936, 48–53.

STOBOROUGH Brit., *Durotriges* (Dorset, GB) 50°40′N, 2°06′W
Pottery kiln, 1st c.
RCHM *Dorset* II part 3 (1970), 592.

STOCKWOOD *cf* Comb End.

STOKE HILL Brit., *Dumnonii* (Devonshire, GB) 50°45′N, 3°31′W
Fortlet. Coin of Carausius (AD 286–93).
TDA XCI (1959), 71 ff: Webster, 1980, 167.

STONE Brit., *Catuvellauni* (Buckinghamshire, GB) 51°48′N, 0°52′W
Pottery kiln. Coins of Vespasian and Domitian.
VCH Buckinghamshire II (1908), 10–11.

STONESFIELD Brit., *Dobunni* (Oxfordshire, GB) 51°51′N, 1°25′W
Villa; hypocausts, mosaics, glass, pottery, coins mostly of 4th c.
VCH Oxfordshire I (1939), 315–16: *Oxon* VI (1941), 1–8: Rainey, 1973, 145–46.

STONEWOOD Brit., *Cantiaci* (Kent, GB) 51°25′N, 0°16′E
Pottery kiln.
VCH Kent III (1932), 131.

STOWTING Brit., *Cantiaci* (Kent, GB) 51°08′N, 1°02′E
Tumulus, diameter at base 24.7 m, height 2.7 m.
Ant XII (1938), 103.

STRAIT OF DOVER *cf* FRETVM GALLICVM

STRATFORD ST MARY Brit., *Trinovantes* (Suffolk, GB) 51°58′N, 0°59′E
AD ANSAM
It. A. 480.3.
Tab. Peut.
PNRB 241.
Minor settlement, between Stratford and Higham on the road between Colchester and VENTA ICENORVM (Caistor St Edmund). Cemetery,

remains of dwellings, site of bridge over the river Stour which here has a bend (*ansa*).
Arch J XXXV (1878), 8: *VCH Suffolk* I (1911), 318: *Britannia* I (1970), 52; VI (1975), 80.

STROUD Brit., ?*Regni* (Hampshire, GB) 51°00′N, 0°58′W
Villa, corridor. Detached bath building. Hypocausts, mosaics, coins AD 270–350.
Arch J LXV (1908), 57–60: LXVI (1909), 33–52.

STUDLEY *cf* Derry Hill

SVESSIONES
Caes. *BG* II, 3, 4, 12, 13; VII, 75; VIII, 6.
Str. IV,3,5 (C. 194); 4,3 (C. 196) (Σουεσσίωνες).
Liv. *Ep* CIV (*Suessones*).
Lucan. I, 423 (*Suessones*).
Plin. *NH* IV, 106 (*Suessiones liberi*).
Ptol. II,9,6 (Οὐέσσονες).
It. A. 362.2, 380.1, 380.6 (*Suessonas*); 379.6 (*Augusta Suessonum*).
Tab. Peut. (*Aug. Suessorum*).
Conc. Agrip. a. 346.
Not. Dig. Occ. IX, 35 (*Suessionensis*).
Not. Gall. VI, 3 (*civitas Suessionum*).
Not. Tiron. LXXXVII, 39 (*Suessio*).
Oros. VI, 7.
Conc. Aurel. a. 511.
Ven. Fort.
Greg. Tur.
CIL XIII.1, 1690 (*Suessioni*); 3204 (*civis Su[essionis]*); 3261 (*Suessio*); notes, p 543; 3528 (*civitatis Sue(ssionum)*); XIII.2, 9158 (milestone of Tongres, *Aug. Suessionum*).
Tribe of Gallia Belgica with capital at AVGVSTA SVESSIONVM (Soissons, not on this map).

SVINDI/VNVM *cf* Le Mans

SVLLONIACIS *cf* Brockley Hill

SURESNES Lug., *Parisii* (Hauts-de-Seine, F) 48°52′N, 2°14′E
Cemetery, 2nd and 3rd c.
Duval, 1961, 231.

SUTTON COURTENAY Brit., *Atrebates* (Oxfordshire, GB) 51°38′N, 1°17′W
Villa, courtyard. Tessellated floors, bricks, tiles, pottery, coins 1st to early 5th c.
JRS LVII (1967), 198.

SWINDON Brit., *Dobunni* (Wiltshire, GB) 51°33′N, 1°48′W
Villa; hypocausts, painted plaster, coins of Constans. Stone quarry.
WANHM L (1942–44), 100: *VCH Wiltshire* I (1957), 112.

SYNDERCOMBE Brit., ?*Dumnonii* (Somerset, GB) 51°04′N, 3°22′W
Iron working; slag, pottery of 1st and 2nd c.
Bull Som Arch I (1956), 4.

TADEN Lug., *Coriosolites* (Côtes du Nord, F) 48°29′N, 2°01′W
At 'Muraille de l'Oeuvre', 'Champ de la Croix-Rouge', a villa; 200 coins of the 3rd c; cement, tegulae, potsherds. At 'Champ des Boissières' and 'La Granville', Gallo-Roman foundations and debris. At 'Trelat', traces of iron smelting.
BSCN V (1867), appendix 96; XLVII (1909), 28.

TAKELEY Brit., *Trinovantes* (Essex, GB) 51°52′N, 0°15′E
Tumulus; bronze vessels, terra sigillata.
VCH Essex III (1963), 185.

TAMAR, river Brit., *Dumnonii* (Devon and Cornwall, GB)
TAMARVS FLVMEN.
Ptol. II,3,3 (Ταμάρου ποταμοῦ ἐκβολαί).
Rav. Cos. 108.26 (*Tamaris*).
PNRB 465.

TAMARVS FLVMEN *cf* river Tamar

TAMESIS FLVMEN *cf* river Thames

TANATVS INSVLA *cf* Thanet

TARRANT HINTON Brit., *Durotriges* (Dorset, GB) 50°54′N, 2°06′W
Villa; mosaics, coins of 3rd and 4th c.
RCHM Dorset IV (1972), 99: *PDNHAS* XCV (1973), 91: Rainey, 1973, 147.

TARVENNA *cf* Thérouanne

TAVERNY Lug., *Parisii* (Val d'Oise, F) 49°02′N, 2°13′E
TABERNIACVM (8th c.).
Probable fort.
Duval, 1961, 233: Roblin, 1971, 17, 64, 122, 159, 189, 245.

TELGRUC Lug., *Osismii* (Finistère, F) 48°14′N, 4°21′W
'le Caon/Pen ar C'haon'; salt manufactory. Terra sigillata from Lezoux, 2nd c.
Gallia XXX (1972), 2, 8, 19.

TERMINIERS Lug., *Carnutes* (Eure-et-Loir, F) 48°05′N, 1°45′E
Gallo-Roman foundations, 'Villaurs'; hypocaust, mosaic, terra sigillata.
Nouel, 6, 16, 26, 36.

TEVCERA *cf* Thièvres

THAMES, river Brit.
TAMESIS (or TAMESA) FLVMEN.
Caes. *BG* V,11,8; V,18,1 (*Tamesis*).
Tac. *Ann* XIV, 32 (*in aestuario Tamesae*).
Ptol. II,3,4 and 11 (Ἰαμήσα).
Cass. Dio XL,3,1; LX,20,5; LX,21,3; LXII,1,2 (Ταμέσα).
Oros. VI,9,6 (*Tamensem*).
Rav. Cos. 106.38 (*Tamese*).
Gildas, 3 and 11.
Beda *Ecc. Hist.* I,2; II,3; III,22; IV,6.
Hist. Brit. II,9,19–20.
PNRB 466.

THANET (Isle of) Brit., *Cantiaci* (Kent, GB) 51°21′N, 1°20′E
TANATVS INSVLA.
Ptol. II,3,14 (Τολιάτις νῆσος).
Solinus, 22, 8 (*Tanatus*).

Rav. Cos. 105.30 (*Taniatide*).
Isidorus XIV,6,3 (*Thanatos*).
PNRB 468.

THATCHAM Brit., *Atrebates* (Berkshire, GB) 51°24′N, 1°16′W
Minor settlement, occupied mid 3rd to late 4th c, on road between Silchester and Cirencester. Floors, wells, pits, pottery, coins.
TNFC VII (1934–37), 219–55.

THÉROUANNE Bel., *Morini* (Pas-de-Calais, F) 50°38′N, 2°15′E
TARVANNA or TARVENNA, later CIVITAS MORINORVM.
Ptol. II,9,4 (Ταρουάννα).
It. A. 376.4, 378.9, 379.1 (*Tarvenna*).
Tab. Peut. (*Tervanna*).
CIL XIII.1, p 560, 3560–61, 8727.
Cf MORINI.
Town, capital of the CIVITAS MORINORVM, road centre. Late-Empire walls, mint, four important cemeteries. Site occupied until destroyed AD 1553.
Blanchet, *Enceintes*, 122: *RN* XLIV (1962), 58–69, 339–356; XLVII (1965), 607; LI (1969), 353–362: *Gallia* XXI (1963), 339; XXIII (1965), 297; XXV (1967), 199: Ringot, 170: Leduque, *Morinie*, 95: *BSAAM* XXI (1971), 468: Delmaire, 1976, 127–188; *Septentrion* X (1980), 41–60.

THEYDON GARNON Brit., *Trinovantes* (Essex, GB) 51°43′N, 0°08′E
Tile kiln.
VCH Essex III (1963), 188.

THIAIS Lug., *Parisii* (Val-de-Marne, F) 48°46′N, 2°23′E
Gallo-Roman foundations, coin hoard 6000 bronze (*c* 276).
Duval, 1961, 232, 247, 278: *Gallia* XVII (1959), 271: *MSNAF* 9th series IV (1968), 19–45.

THIEPVAL Bel., *Ambiani* (Somme, F) 50°03′N, 2°42′E
Roman camp. In the marsh, many Gallo-Roman antiquities.
Leduque, *Ambianie*, 30.

THIEULLOY-L'ABBAYE Bel., *Ambiani* (Somme, F) 49°49′N, 1°56′E
Villa.
Agache, fig. 428: Vasselle, 330; Agache-Bréart, 123.

THIÈVRES-SUR-AUTHIE Bel., *Ambiani* (Somme, F) 50°07′N, 2°27′E
TEVCERA/TEVARA.
Tab. Peut.
Major settlement on the road between Amiens and Arras, at the border between the *Ambiani* and the *Atrebates*. Sarcophagi, coin hoard Gordian-Gallienus.
de Loisne, A., *Dictionnaire topographique du département du Pas-de-Calais*, Paris, 1907, 168: Vasselle, 330: Leduque, *Ambianie*, 29, 134.

THIVERNY Bel., *Bellovaci* (Oise, F) 49°15′N, 2°26′E

Sanctuary near the road between Senlis and Beauvais, destroyed 1st c, rebuilt in the Antonine period; pool at spring; habitat 700 m to the south, stone quarry, pits, burials.
Ogam XVII (1965), 278–94: *Gallia* IX (1951),82; XII (1954), 145; XV (1957), 168.

THORNBOROUGH *cf* Bourton Grounds

TIDBURY RING Brit., *Belgae* (Hampshire, GB) 51°11′N, 1°20′W
Villa; inside Iron Age hill fort. Bricks, tiles, plaster, coins of Constantine I.
PHFC XVI (1944–46), 38–39: *JRS* XLIII (1953), 94.

TILBURY Brit., *Trinovantes* (Essex, GB) 51°28′N, 0°22′E
Three pottery kilns, 1st c.
MPBW Arch Exc, 1969, 8.

TILFORD Brit., *Regni* (Surrey, GB) 51°11′N, 0°44′W
Pottery kilns; 4th c.
VCH Surrey IV (1912), 362: *SyAC* LI (1950), 29–56.

TILLINGHAM Brit., *Trinovantes* (Essex, GB) 51°41′N, 0°53′E
Salt workings, briquetage.
VCH Essex III (1963), 191.

*****TILLOLOY** Bel., *Ambiani* (Somme, F) 49°38′N, 2°45′E
Gallo-Roman foundations, hypocaust.
Agache-Bréart, 1975, 123: Agache, 1978, 262, 267, 311.

TILLOY-FLORIVILLE Bel., *Ambiani* (Somme, F) 49°59′N, 1°37′E
Vicus, temple nearby, 'Vers Baillon'.
Agache, 1972, 323; 1978, 392.

*****TILLOY-LÈS-CONTY** Bel., *Ambiani* (Somme, F) 49°45′N, 2°11′E
Villa at 'Les Marlis'.
Vasselle, 330: Agache-Bréart, 1975, 124.

TILLOY-LÈS-HERMAVILLE Bel., *Atrebates* (Pas-de-Calais, F) 50°20′N, 2°33′E
Gallo-Roman foundations.
Jelski, 143.

TINTAGEL Brit., *Dumnonii* (Cornwall, GB) 50°40′N, 4°45′W
RIB 2231.
Milestone, of Licinius (AD 308–24).
Sedgley, 1975, 22–3.

TITSEY Brit., ?*Atrebates/Cantiaci* (Surrey, GB) 51°16′N, 0°01′E
Temple in temenos, *c* AD 100–260. Corridor villa, hypocausts, baths.
Lewis, 1966, *passim*; *SyAC* XLIV (1936), 84–101: *VCH Surrey* IV (1912), 367–9.

TOCKINGTON Brit., *Dobunni* (Avon, GB) 51°34′N, 2°32′W
Villa, courtyard. Hypocausts, mosaics 4th c.
TBGAS XII (1888), 159–69; XIII (1889), 196–204: Rainey, 1973, 149.

TOPSHAM Brit., *Dumnonii* (Devonshire, GB) 50°41′N, 3°29′W
Major settlement; port of Exeter, probably originating in mid-1st c supply-base; occupied into late 4th c. Foundations, slates, tiles, pottery, glass, coins etc.
PDAS III (1937–38), 4–23, 67–82; IV (1948), 20–23; XXXIII (1975), 209–265.

TÔTES Lug., *Caletes* (Seine-Maritime, F) 49°41′N, 1°03′E
Temple, 'Butte des Buits, forêt de Bord'.
Coutil II (*Louviers*), 233: Vesly, 25: Deglatigny, 1925, 35.

TOTTERNHOE Brit., *Catuvellauni* (Bedfordshire, GB) 51°53′N, 0°34′W
Villa, courtyard. Baths, hypocausts, tessellated floors, pottery mostly of 4th c.
JRS XLVII (1957), 214–15: Matthews, C. L., *Ancient Dunstable*, Dunstable, 1963, 61–64: Rainey, 1973, 149.

TOURLAVILLE Lug., *Unelli* (Manche, F) 49°38′N, 1°34′W
Possible fort at 'les Fosses Câtel'; foundations, sherds, querns, bricks, tegulae, statuettes in terra cotta, a figurine in bronze, gold ornaments. Two coin hoards, one of the early Empire, the other of the 4th c.
de Gerville, 1854, 207: *BSAN* LIII (1955–6), 197–263.

TOURNELLES (Les) Bel., *Suessiones* (Oise, F) 49°22′N, 2°53′E
Temple in the Forest of Compiègne.
Grenier, *Manuel* IV.1 (1960), 327.

TOURNOISIS Lug., *Carnutes* (Loiret, F) 48°00′N, 1°38′E
Gallo-Roman foundations, quern.
Nouel, 30, 31.

TOUROUVRE Lug., *Esuvii* or *Sagii* (Orne, F) 48°35′N, 0°40′E
Minor settlement? Bricks, tiles, slag.
MSAN V (1829–30), 90–120.

TOURS-EN-VIMEU Bel., *Ambiani* (Somme, F) 50°02′N, 1°41′E
Temple, other Gallo-Roman foundations, bust of Cybele, terra sigillata, coins.
BSAP, 1852, 41; 1859, 14: Vasselle, 330: Leduque, *Ambianie*, 164, n. 727.

TOVIVS FLVMEN *cf* river Tywi

TRECASTLE HILL Brit., *Silures* (Dyfed, GB) 51°58′N, 3°43′W
RIB 2260, 2261.
Milestone of Postumus, AD 258–68.
Nash-Williams, 1969, 185 no 7, 186 no 11; Sedgley, 1975, 32.

TRÉGRON Lug., *Osismii* (Côtes du Nord, F) 48°36′N, 3°24′W
At 'la Croix des Écuries' and 'Kerjefiou' Gallo-Roman foundations and debris.
MSCN I (1883–84), 322.

TRÉGUEUX Lug., *Coriosolites* (Côtes du Nord, F) 48°29′N, 2°44′W
'Preauren', Gallo-Roman foundations extending for more than one hectare. At 'Sainte Marie' foundations, burials, an uninscribed milestone, a 'barbaric' statue now in the museum at Saint-Brieuc.
MSCN I (1883–84), 174.

TRÉHORENTEUC Lug., *Veneti* (Morbihan, F) 48°01′N, 2°17′W
'Bourg à la Mazeraie', villa; hypocaust, 28 brick pillars; tegulae, pottery, glass, slag, bronze.
BSPM, 1907, 284; 1927, 78–79: *Ann Bret* LX (1953), 415.

TRELOY Brit., *Dumnonii* (Cornwall, GB) 50°25′N, 5°01′W
Tin working; brooches, coins.
VCH Cornwall I (1906), 371.

TREMBLAY *cf* Verneuil-en-Halatte

TRÉPORT (LE) Bel., *Ambiani* (Seine-Maritime, F) 50°04′N, 1°22′E
VLTERIOR PORTVS
?Caes. *BG* IV, 23.
Ultriportensis (12th c).
Port; sarcophagi, tiles, pottery.
Agache, fig. 98: Leduque, *Ambianie*, 163.

TRESSAINT Lug., *Coriosolites* (Côtes du Nord, F) 48°26′N, 2°02′W
Gallo-Roman foundations; probable bath-house on the banks of the Rance; foundations, tegulae, cement near the town; at 'Puits Harel' traces of a hypocaust and other remains; at 'la Mercerie' a pot containing 280 Roman coins, tegulae.
BSCN XLVII (1909), 21–3: *Ann Bret* LXIV (1957), 97–100.

TRETHEVEY Brit., *Dumnonii* (Cornwall, GB) 50°40′N, 4°43′W
RIB 2230.
Milestone of Gallus and Volusian (AD 251–53).
Sedgley, 1975, 22.

TRÉVRON Lug., *Coriosolites* (Côtes du Nord, F) 48°23′N, 2°04′W
'la Basse-Landrie', 'Maumusson', Gallo-Roman foundations, tegulae, potsherds, cement.
BSCN XLVII (1909), 28, 29.

TRIGAVOU Lug., *Coriosolites* (Côtes du Nord, F) 48°32′N, 2°05′W
Christianised milestone.
BSCN XLVII (1909), 85–86.

TRINOVANTES
Caes. *BG* V, 20, 21, 22 (*Trinobantes*).
Tac. *Ann.* XIV, 31 (*Trinobantibus*).
Ptol. II,3,11; II,3,14 (Τρινόαντες).
Oros. VI,9 (*Trinobantum*).
Beda, *Eccl. Hist* I, 2 (*Trinouantum*).
PNRB 475–6.
British tribe occupying Essex and southern Suffolk. Allies of Caesar, who describes them as *prope firmissima earum regionum civitas*. Subjected by Cunobelinus of the *Catuvellauni* (*qv*) c AD 5–10; re-established by the Romans as a *civitas* soon after

the conquest. It is uncertain whether the capital of this was at CAESAROMAGVS (Chelmsford, *qv*) or CAMVLODVNVM (Colchester, *qv*).

Rivet, 1964, 162–3; Dunnett, 1975, *passim*: Wacher, 1975, 195–8: Rodwell-Rowley, 1975, 85–101: Frere, 1978, 52–3, 57, 60–61, etc.

TRISANTONA FLVMEN *cf* river Arun

TROSLY-BREUIL Bel., *Suessiones* (Oise, F) 49°24'N, 2°58'E
Stone quarries; at 'La terre à carreaux', mosaics.
Gallia Sup. X.1.1 (1957), no 82.

TRUCKLE HILL *cf* North Wraxall

TWYFORD Brit., *Belgae* (Hampshire, GB) 51°01'N, 1°19'W
Villa, winged corridor, built mid-2nd c. Baths, hypocausts, mosaic; coin of Constantine I.
JRS XLIX (1959), 131: Rainey, 1973, 149.

TWYN-Y-BRIDDALLT Brit., *Silures* (Mid Glamorgan, GB) 51°40'N, 3°26'W
Temporary camp.
JRS XLIX (1959), 102. RCAHM (Wales), *Glamorgan* I, ii, 1976, 99–102.

TYWI or **TOWY**, river Brit., *Demetae* (Dyfed, GB)
TOVIVS FLVMEN
Ptol. II,3,2 (Τουβίου ποταμοῦ ἐκβολαί).
PNRB 474.

VGGADE *cf* Caudebec-lès-Elbeuf

VLTERIOR PORTVS *cf* le Tréport

VNELLI
Caes. *BG* II, 34; III, 7, 11, 17; VII, 75.
Plin. *NH* IV, 107 (*Venelli*).
Ptol. II,8,2 and 5 (Οὐενέλλοι).
Cass. Dio XXXIX, 45 (Οὐενέλλους).
CIL XIII.1, pp 494–5.
Tribe of Lugdunensis, occupying the Cotentin. Replaced under the late Empire by the CIVITAS CONSTANTIA (in Lugdunensis II) with capital at Coutances (*qv*).

ULEY Brit., *Dobunni* (Gloucestershire, GB) 51°42'N, 2°13'W
At West Hill (N of Uley Bury hill-fort) pre-Roman shrine succeeded by first a timber then a stone (4th c.) Romano-Celtic temple demolished *c.* AD 400 and replaced by domestic buildings. Finds incl. Dobunnic, Durotrigan and Roman coins (Trajan-Theodosius) pottery 1st to 5th c., altar to and statue and figurines of Mercury, and many *defixiones*.
RCHM, *Gloucestershire Cotswolds*, 1976, 121–3; Ellison, A., *Excavations at West Hill, Uley, 1977*, Bristol, 1978: *Britannia* IX (1978), 457; X (1979), 323, 340–5, 349; XI (1980), 385–7, 411–4; XII (1981), 370; McWhirr, 1981, 156–8.

UPCHURCH Brit., *Cantiaci* (Kent, GB) 51°23'N, 0°39'E
Many pottery kilns, 1st c onwards.
Arch Cant LXXVII (1962), 190–95: *VCH Kent* III (1932), 132–4.

UP MARDEN Brit., *Regni* (West Sussex, GB) 50°54'N, 0°52'W
Villa; corridor. Baths, outbuildings. Occupied early 3rd to late 4th c.
JRS LVII (1967), 198; LVIII (1968), 202; LIX (1969), 231. Down, A., *Chichester Excavations* IV, 1979.

USHANT *cf* Île d'Ouessant

USK Brit., *Silures* (Gwent, GB) 51°42'N, 2°54'W
BVRRIVM.
Ptol. II,3,12 (Βούλλαιον).
It. A. 484.5, 485.1 (*Burrio*).
RIB 396.
PNRB 285.
Fort, fortress, major settlement. Neronian fortress for Legio XX, *c* 19.5 ha, *c.* AD 55–67. Flavian fort, abandoned mid-2nd c., civil settlement until *c* 350. Granaries, baths, etc.
Manning, W. H., *Report on the Excavations at Usk 1965–76: the Fortress Excavations 1968–71*, Cardiff, 1981.

USK, river Brit., *Silures* (Gwent, GB)
ISCA FLVMEN.
Rav. Cos. 108.27.
PNRB 378.

VXELA FLVMEN
Ptol. II,3,2 (Οὐεξάλλα εἴσχυσις).
PNRB 482–3.
It is uncertain whether this refers to the river Axe or to the river Parrett (both in Somerset, GB).

VXISAMA or **VXANTIS INSVLA** *cf* Île d'Ouessant

VAGNIACIS *cf* ?Springhead

VAGORITVM *cf* ?Bazouge-de-Chémeré (La)

VALCANVILLE Lug., *Unelli* (Manche, F) 49°38'N, 1°20'W
Pottery kiln on the road between Valognes and Barfleur. Tegulae, querns, coins of the early Empire, one gold piece of Lucius Verus: 1st and 2nd c.
de Gerville, 1854, 213: *BSAIC* XXIV (1900), 85, 100: *BSAN* LIII (1955–56), 197–263.

VALLANGOUJARD Lug., *Veliocasses* (Val-d'Oise, F) 49°08'N, 2°07'E
Gallo-Roman foundations of buildings of many kinds, early-3rd to end-4th c; coins.
BAVF I (1965), 70–78: *Gallia* XVII (1959), 276.

VALMONT Lug., *Caletes* (Seine-Maritime, F) 49°44'N, 0°31'E
Cremation cemetery.
Gallia XII (1959), 336.

VALOGNES Lug., *Unelli* (Manche, F) 49°30'N, 1°28'W
ALAVNA.
It. A. 386.6.
Tab. Peut.
Town, 'Alleaume', 'Le Castelet', on the road between Carentan and Cherbourg. Occupied 1st to 5th c. Forum, theatre, baths, aqueducts, temple of

Victory, pits, foundations. Coin hoard from Julia Domna to Postumus, another of end of 3rd c; plaques of bronze, fibulae, candelabra, bronze pins. Inhumation of 3rd c on the farm 'de la Côeffe'. Terra sigillata IRIDVRNO, LHOSCRI (on a lamp); glass. Chalk quarries; possible mint; bricks, querns, architectural fragments.

MSAN XIV (1845), 317–31: de Gerville, 1854, 79, 214–6; *Ann Manche*, 1867–1874: *Mém Soc Val* III (1882–4), 115–23: *BSAIC* XXIV (1900), 101–2; supp. XXX (1908), 100: *BSNÉP* XIII (1905), 142–8: *BSAN* XXIX (1913), 230; LIII (1955–6), 197–263; LIV (1957–8), 8th June; LV (1959–60), 5th Dec: *Ann 5 Norm* LXI (1953), 19–20: Grenier, *Manuel* III.2 (1958), 959–63; IV.1 (1960), 352–3: Bouhier, 1962, *v* Valognes.

VATON *cf* Falaise

VATTEVILLE-LA-RUE Lug., *Lexovii* (Seine-Maritime, F) 49°30′N, 0°41′E
Temple.
Deglatigny, 1927, 35: *BCASI* XVI, 129–39.

VAUCHELLES-LÈS-DOMART Bel., *Ambiani* (Somme, F) 50°03′N, 2°03′E
Gallo-Roman foundations.
Agache, fig. 173, 174: Agache-Bréart, 1975, 126: Agache, 1978, 40, 115, 158, 186, 231.

VAULX-VRAUCOURT Bel., *Atrebates* (Pas-de-Calais, F) 50°08′N, 2°55′E
Minor settlement; pottery.
Agache, fig. 483: *BCDMHPC* IX.1 (1971), 48: Agache-Bréart, 1975, 141: Agache, 1978, 270.

*VAUX-EN-AMIÉNOIS Bel., *Ambiani* (Somme, F) 49°58′N, 2°15′E
Gallo-Roman foundations: 'Le Sepulchre', sarcophagus.
Agache, fig. 612: Leduque, *Ambianie*, 129: Agache-Bréart, 1975, 126.

VECTIS INSVLA *cf* Isle of Wight

VEL(L)IOCASSES
Caes. *BG* II, 4; VII, 75; VIII, 7.
Plin. *NH* IV, 107.
Ptol. II,8,7 (Οὐελιοκάσιοι).
Not. Tiron. LXXXVII, 58 (*Viliocassus*).
Oros. VI, 7 and 11 (*Velocasses*).
CIL XIII.1, 1717 (*cives* (read *civis*) *Velioca*(*ssis*)); 1998 (*ex civitate Veliocassium*); notes, p 512.
Britannia VIII (1977), 430 (*Veliocas*[*s*]*ium*).

Tribe of Gallia Belgica, then Lugdunensis, occupying the Vexin. Oppida: Bouquelon, Caudebec, Orival, St-Nicolas-de-la-Taille, St-Pierre-de-Varangeville, St-Samson-de-la-Roque, Sandouville, Vernon. Capital: RATOMAGVS or ROTOMAGVS (Rouen, *qv*). Under the late Empire amalgamated with the *Caletes* (*qv*) to form the CIVITAS ROTOMAGENSIVM in Lugdunensis II, still with its capital at Rouen.

Mangard, M., *La tombe gallo-romaine chez les Calètes et les V.*, Rouen: Saforge, J.-P., Les V. et les Calètes, *BSNÉP*, 1959: *BSNAF*, 1969, 294–320: P-W *Supp*. XV (1978), 777–83.

VELLAVNODVNVM
Caes. *BG* VII, 11 and 14.
Oppidum of the *Senones*.
Cf Château-Landon, Girolles.

*VENDEUIL-CAPLY Bel., *Bellovaci* (Oise, F) 49°37′N, 2°18′E
?BRATVSPANTIVM.
Caes. *BG* II, 13.

Major settlement. Colline du Calmont, oppidum *c* 125 ha, perhaps *Bratuspantium*; numerous La Tène works and Gallic coins. Colline du Catelet, early Roman fort (?of Crassus), earliest constructions of stone, system of level military terraces, temple. Augustan settlement (much Arretine ware) of regular plan, abandoned *c* 260. Two theatres; one on the hillside (diameter 73 m, second half 1st c to 2nd c?), one in the valley of St-Denis (diameter 83 m, early 1st c, reconstructed 2nd c). Rural frontier sanctuary; road junction. Cellar, pits, mosaics, small finds.

Cambry, J., *Statistique du département de l'Oise*, 1803, 217: Mouret, *Histoire de Breteuil*, 1821: *Bull Mon*, 1844–5, 41–6: Baticle, A., *Nouvelle histoire de Breteuil*, Beauvais, 1891: Leblond, V., *L'oppidum Bratuspantium des Bellovaques*, Beauvais, 1909: *Gallia, Sup.* X.1.1 (1957), nos 84–5; XIX (1961), 305; XXI (1963), 372; XXIII (1965), 322; XXV (1967), 199; XXIX (1971), 228: Duval, 1961, 88–90: *BSPF*, 1962, 357: *Celticum* VI (1962), 201–13; IX (1963), 229–239; XV (1966), 144: Agache, fig. 70–74, 242, 381–6, 585–91: *RN* XLIV (1962), 334: *RIO*, March 1963, 1–24, 161–8: *Mémoires de Photo-Interprétation* IV (1967), 10–18: *BSAP*, 1972, 323: *Bulletin de la Société française de Photogrammetrie* no 5, April 1972: Leduque, *Ambianie*, 150: Montfaucon, 1972, V.88 part 1, LII: Agache-Bréart, 1975, 144, 151: Roblin, 1978, 232–4.

VENETI
Caes. *BG* II, 34; III, 7–9, 11, 16–18; VII, 75.
Liv. *Ep* CIV.
Str. IV,4,1 (C.194) (Οὐένετοι).
Plin. *NH* IV, 107 and 109.
Flor. I, 45 (III, 10).
Ptol. II,8,6 and 7.
Cass. Dio XXXIX, 40.
Tab. Peut.
Not. Dig. Occ. XXXVII, 5 and 16 (*praefectus militum Maurorum Benetorum, Benetis*).
Not. Gall. III, 8 (*civitas Venetum*).
Not. Tiron. LXXXVII, 71, n 93 Z.
Oros. VI, 8.
Conc. Aurel. a. 511.
Ven. Fort. *Vita S Albini* x, xv.
Greg. Tur. *HF* IV, 4; V, 19 (26); V, 22 (29); VIII, 25; IX, 8; X, 9.
CIL XIII.1, 1709 (*Veneto*); 2950 (*curator r*(*ei*) *p*(*ublicae*) *civit*(*atis*) *Venet*(*orum*)); notes, p 489.

Armorican tribe included in Lugdunensis (later Lugdunensis III) with capital at DARIORITVM, later known as VENETIS (Vannes, not on this map).
P-W VIII A.1 (1955), 705–86.

VENN BRIDGE Brit., *Durotriges* (Somerset, GB) 50°58′N, 2°45′W

RIB 2229.
Milestone, inscription of Flavius Valerius Severus (AD 305–06).
Sedgley, 1975, 22.

VENTA BELGARVM *cf* Winchester

VENTA SILVRVM *cf* Caerwent

VENTES (Les) Lug., *Aulerci Eburovices* (Eure, F) 48°57′N, 1°05′E
Pottery kiln, end-1st to early-2nd c.
Gallia XX (1962), 424.

VER-LÈS-CHARTRES *cf* Chartres

VERLVCIO *cf* Sandy Lane

VERNEUIL-EN-HALATTE Bel., *Silvanectes* (Oise, F) 49°17′N, 2°31′E
'Bufosse', villa; mosaics, hypocaust, bath, end of 1st to end of 2nd c. Tremblay, oppidum. 'La Cavée Douche', Gallo-Roman occupation. At Montlaville, two further villas, 'les Tronces' and 'les 18 arpents'.
Mém Soc Clermont: *Gallia* IX (1951), 82; XII (1954), 144; *Sup* X.1.1 (1957), 59: *RN* XXXVIII (1956), 289–306: *RIO* XVI (1964), 185–205.

VERNON Lug., *Aulerci Eburovices* (Eure, F) 49°05′N, 1°29′E
Gallo-Roman cemetery. Camp de Vernonnet, 'Camp de César, camp romain, Camp de Mortagne'; oppidum on the frontier between the *Eburovices* and the *Veliocasses*. Musée A.-G. Poulain.
BSPF XI (1914), list XXIX, 155: Duval, 1961, 66: *Gallia* XXIV (1966), 262.

*****VERS-SUR-SELLES** Bel., *Ambiani* (Somme, F) 49°50′N, 2°14′E
Probable fort. ?Villa; mosaics, burials, pottery, coins.
Josse, H., *Notice sur V.*, 1880: *Gallia Sup* X.1.1 (1957), 97; XXIII (1965), 310: Agache, fig. 168: Vasselle, 331: Leduque, *Ambianie*, 153, n 502–6, 510: Agache-Bréart, 1975, 127.

VERT-EN-DROUAIS Lug., *Carnutes* (Eure-et-Loir, F) 48°46′N, 1°18′E
Gallo-Roman foundations.
Gallia XX (1972), 316.

VERVLAMIVM *cf* St Albans

VICINONIA FLVMEN *cf* river Vilaine

VICVS DOLVCENSIS *cf* ?Halinghen, ?Isques

VIDVCASSES
Plin. *NH* IV, 107.
Ptol. II,8,2 and 5 (Βιδουκάσιοι).
Not. Tiron. LXXXVII, 64 (*Vidiocasus*).
CIL XIII.1, p 496 (notes), 3162 ('Le Marbre de Thorigny': *civitatis Viducass(ium)*), 3166 (*C(ivitatis) V(iducassium)*).
Tribe of Lugdunensis. Oppida: Moult, Soumont-St-Quentin. Capital: AREGENVAE (Vieux, *qv*). Under the late Empire amalgamated with the CIVITAS BAIOCASSIVM, and thus not listed in *Not. Gall.*
P-W *Supp.* XV (1978), 914–7.

VIEIL-ÉVREUX (Le) Lug., *Aulerci Eburovices* (Eure, F) 49°00′N, 1°14′E
?GISACVM (*qv*).
?MEDIOLANVM AVLERCORVM (*cf* Évreux).
CIL XIII.1, p 510; 3197, 3202, 3206, 3208.
Town and pre-Roman road junction, capital of the *Aulerci Eburovices* until succeeded by Évreux end of 1st c; religious centre active until 3rd c, burned 276. 'Basilica' (?temple, praetorium or palace), theatre, baths, gymnasium, aqueduct (Coulonges, Le Misérey), pits. At Cracouville, Romano-Celtic temple succeeding pre-Roman one of similar design.
Rever, F., *Mémoire sur les ruines du V.E.*, Paris-Rouen, 1827: *BSFFA* III (1913), 80: Hermier, A., *La ville gallo-romaine de Mediolanum Aulercorum*, Évreux, 1936: *Gallia* II (1943), 191–206; VII (1949), 122; XII (1954), 345; XX (1962), 423: *Ann Norm* VIII (1958), 400: Grenier, *Manuel* III.2 (1958), 953–4; IV.1 (1960), 191–9, 342–4; IV.2 (1960), 757–68: Duval, 1961, 140, 150, 178, 187: *BSNÉP* XXXVIII (1965), 188.

VIENNE Lug., *Baiocasses* (Calvados, F) 49°17′N, 0°36′W
Gallo-Roman foundations, 'Mesnils', 'les Petits Boraux', on the road between Bayeux and le Bac du Port. Coins of the 3rd c; red and grey coarse pottery, some terra sigillata including COSMIAN M.
de Caumont, 1831, 234; 1875, 522.

VIERVILLE Lug., *Carnutes* (Eure-et-Loir, F) 48°23′N, 1°55′E
Gallo-Roman foundations, 'La Croix Butte'.
Jalmain, air photograph.

VIEUX Lug., *Viducasses* (Calvados, F) 49°06′N, 0°26′W
AREGENVAE
Ptol. II,8,2 ('Αρηγενούα).
Tab. Peut. (*Araegenue*).
CIL XIII, 3162–3176.
Wuilleumier 341.
Cf VIDVCASSES.
Town, capital under the early Empire of the CIVITAS VIDVCASSIVM, on the road between Bayeux and Jublains. Baths, theatre, aqueducts, pits, houses, forum, temple, public monuments. Potter's kiln, bakery, oculist's stamp. Coins of the *Baiocasses* and *Lexovii*, more than 1000 coins from 19 to 388; one coin inscribed VERGOBRET, one with Christian symbol (5th c), one African coin, one of Diadumenian. Terra sigillata.
MSAN I.2 (1824), 472–89; III (1826), 127–77; IV (1829), 35–7; VI (1833), 319–61; X (1837), 683–85; XII (1841), 342–54; XIII (1844), 300–3; XIX (1851), 37; XX (1855), 458–85; XXVIII (1873), 96–7: de Caumont, 1831, 129–43; 1846, 131–43; 1854, 548–56; 1864, 825–54: Charma, A., *Archéologie*, Paris, 1863, 131–44: *BSAN* III (1864), 292–3; IV (1867), 254, 292; VII (1875), 86; IX (1881), 233–8, 241–8; XXVIII (1913), 239–40; XXXII (1917), 360; XXXV (1924), 398–407, 483–6; XXXVIII (1930), 468–75; LII (1953–4), 292–308; LIII (1956), 394; LIV (1959), 422: Desjardins, *Géo. Gaule* III (1885), 197–211: Blanchet, 1905, 320–1: *BSNÉP* XIII (1905),

186–91: *MSNAF* LXIV (1909), 225–335: *CRCAF*, 1909, 502–15: *CRAI*, 1910, 559–63: Espérandieu III (1911), 3040–44: Pflaum, H. G., *Le Marbre de Thorigny*, Paris, 1948, *passim*: Grenier, *Manuel* III.2 (1958), 913–6; IV.1 (1960), 350–5: *Ann Norm* X (1960), 3–24: *Gallia* XXVIII (1970), 272.

VIEUX-BOURG (Le) Lug., *Osismii* (Côtes du Nord, F) 48°23'N, 3°00'W
'Bourg-Blanc', 'Clos du Vieux-chastel', 'Collédic', 'Coz-Chausée', Gallo-Roman foundations and debris. At 'Parc-Pilate' silver and bronze coins, potsherds, amphora sherds, loom weights.
MSCN I (1883–84), 261.

VIEUX-MOULIN Bel., *Suessiones* (Oise, F) 49°23'N, 2°56'E
Quarry. St-Pierre-en-Chastre (forêt de Compiègne). Roman camp? Oppidum of the Suessiones.
DAG: Jullian, *HG* III (1909), 549, n 1: *BSPF* XIV (1917), list lxi, 469: Grenier, *Manuel* I (1931), 194: *Gallia* VII (1949), 114.

VIEUX-ROUEN-SUR-BRESLE *Caletes/Ambiani* (Seine-Maritime, F) 49°50'N, 1°43'E
Temple, beneath the church; tiles, coins. At the hamlet of Brétuzel, tower, statue, coins.
Leduque, *Ambianie*, 164, n. 709–11: Agache, 1978, 47, 270, 296, 297, 298, 299, 303, 444, 446, 466.

VILAINE, river Lug., *Redones/Veneti* (Ille-et-Vilaine/Morbihan, F)
VICINONIA FLVMEN
?Ptol. II.8,1 (Ἡρίου ποταμοῦ ἐκβολαί).
Greg. Tur. *Hist Francorum* V, 19; X, 9.

VILLE DES GAULES *cf* Pierrefonds

VILLEJUIF Lug., *Parisii* (Val-de-Marne, F) 48°48'N, 2°22'E
Gallo-Roman foundations, tiles, bones.
CVP, 1909, 118: Toussaint, *Seine*, 29: Duval, 1961, 32, 36, 55, 227, 283: Roblin, 1971, 81, 185, 277.

VILLENEUVE-LE-ROI Lug., *Parisii* (Val-de-Marne, F) 48°44'N, 2°25'E
Gallo-Roman foundations, early Empire, over pre-Roman dwellings; cremation cemetery, Christian sarcophagus.
Duval, 1961, 229: Roblin, 1971, 81, 116, 286.

VILLENEUVE-ST-GEORGES Lug., *Parisii* (Val-de-Marne, F) 48°44'N, 2°27'E
Gallo-Roman foundations, early Empire, coins.
Duval, 1961, 227–8: Roblin, 1971, 18, 81, 111.

VILLERS-AU-BOIS Bel., *Atrebates* (Pas-de-Calais, F) 50°22'N, 2°40'E
Gallo-Roman foundations.
Jelski, 143.

VILLERS-AU-FLOS Bel., *Atrebates* (Pas-de-Calais, F) 50°05'N, 2°55'E
Gallo-Roman foundations, 'Le Paradis'.
BCDMHPC IX.1 (1971), 49: Agache-Bréart, 1975, 14.

*VILLERS-BOCAGE Bel., *Ambiani* (Somme, F) 49°59'N, 2°20'E
Villa; tiles, coins.
Agache, fig. 247, 457, 486, 491, 606: Agache-Bréart, 1975, 128: Agache, 1978, 63, 71, 114, 186.

*VILLERS-BRETONNEUX Bel., *Ambiani* (Somme, F) 49°52'N, 2°31'E
Gallo-Roman foundations, cellars, coins.
Boulianes, J., *Histoire de V.-B.*, 31: Agache, fig. 196, 208, 215, 216, 224, 499: Vasselle, 331: Leduque, *Ambianie*, 137, n 204: Agache-Bréart, 1975, 128.

*VILLERS-LÈS-ROYE Bel., *Ambiani* (Somme, F) 49°42'N, 2°44'E
Villa. Late *castellum* of 'Vieux-Catil'; weapons, coins, pottery.
Agache, fig. 514: Leduque, *Ambianie*, 143: Agache-Bréart, 1975, 128: Agache, 1978, 215, 246.

VILLETTE (La) Lug., *Viducasses* (Calvados, F) 48°54'N, 0°32'W
Small villa, foundations and small finds extending over 1,5 km.
BSAN VI (1874), 296–98.

VILLIERS-LE-SEC Lug., *Baiocasses* (Calvados, F) 49°17'N, 0°33'W
Small villa beside the road between Bayeux and le Bac du Port. Foundations and small finds.
de Caumont, 1831, 229; 1857, 534.

VILLIERS-SOUS-MORTAGNE Lug., *Esuvii* or *Sagii* (Orne, F) 48°32'N, 0°36'E
Villa. Partially excavated in 1880: hypocaust, baths, mosaic.
BSAN XI (1881–82), 518–547: *BSHAO* LXXIX (1967), 17–39.

VIMY Bel., *Atrebates* (Pas-de-Calais, F) 50°22'N, 2°49'E
Villa, quarry.
Dérolez, 514.

VINDEFONTAINE Lug., *Unelli* (Manche, F) 49°20'N, 1°25'W
Pottery and tile kilns at 'les Michelleries'; coarse pottery of 3rd c.
Gallia XXVII (1970), 274–75.

VINDI/VNVM *cf* Le Mans

VINDOCLADIA *cf* Badbury, Dorset

VIROMANDVI
Caes. *BG* II, 4, 16, 23.
Liv. *Ep* CIV.
Plin. *NH* IV, 106 (*Veromandui*).
Ptol. II,9,6 (Ῥομάνδυες).
It. *A*. 379.4 (*Augusta Veromandorum*).
Tab. Peut. (*Aug. Viromuduorum*).
Not. Gall. VI, 5 (*civitas Veromandorum*).
Oros. VI, 7 (*Veromandi*).
CIL XIII.1, 1465 (*Viromanduo*), 1688 (*Viromand(uo)*); notes, p 556; 3528 (*Civit(atis) Vi(romanduorum)*); XIII.2, 8341, 8342 (*civi Viromanduo*).
Tribe of Gallia Belgica, occupying the Vermandois

with capital at AVGVSTA VIROMANDVORVM (St-Quentin, not on this map).
P-W XVII (1961), 241 sq.

VISMES-AU-VAL Bel., *Ambiani* (Somme, F) 50°01′N, 1°40′E
Gallo-Roman foundations, ?*mansio*; coins.
Vasselle, 331: Agache-Bréart, 1975, 129: Agache, 1978, 438, 452.

VITTEFLEUR Lug., *Caletes* (Seine-Maritime, F) 49°49′N, 0°39′E
Villa (and Frankish tombs).
Cochet, *S.I.*, 457.

VITRY-EN-ARTOIS Bel., *Atrebates* (Pas-de-Calais, F) 50°20′N, 2°59′E
Gallo-Roman foundations; hypocaust.
BCDMHPC IX.2 (1972), 97: Dérolez, 521: Jelski, 139: Agache-Bréart, 1975, 141.

VORGIVM *cf* Carhaix

*****VOYENNES** Bel., *Ambiani* (Somme, F) 49°46′N, 2°59′E.
Villa; hypocaust.
Agache, fig. 198: Agache-Bréart, 1975, 129: Agache, 1978, 257, 260, 262, 263, 267, 380.

VRON Bel., *Ambiani* (Somme, F) 50°19′N, 1°45′E
Temple, 'Bois d'Avesnes', ditch and enclosure; small finds. Gallo-Roman cemetery, later Merovingian.
Agache, fig. 368, 398, 607; *BSAP*, 1972, 324: *Gallia* XXIX (1971), 233: Leduque, *Ambianie*, 175: Vasselle, 332; Agache-Bréart, 1975, 129: Agache, 1978, 54, 184, 393, 400, 414, 428, 455.

WACQUEMOULIN Bel., *Bellovaci* (Oise, F) 49°30′N, 2°37′E
Temple, 'Les 30 Mines'.
Agache, 1972, 324: Agache-Bréart, 1975, 152: Agache, 1978, 394.

WADDON HILL Brit., *Durotriges* (Dorset, GB) 50°48′N, 2°47′W
Fort (situated in Iron Age hill fort) constructed under Claudius, earth rampart, timber buildings, soon abandoned. Small finds, some in Bridport museum, include bronze tricorn bull and bronze tankard handle, indicating importance of the site in pre-Roman times.
PDNHAS LXXXII (1960), 88–108; LXXXVI (1964), 135–49; XCI (1969), 180; CI (1979), 51–90.

WADEFORD *cf* Combe St Nicholas

WADFIELD Brit., *Dobunni* (Gloucestershire, GB) 51°56′N, 1°58′W
Villa, courtyard. Baths; mosaic pavement (roofed over on site); outbuilding. ?4th c.
JBAA ns I (1895), 242–50: *TBGAS* XC (1971), 124–28: Rainey, 1973, 150.

WAKERING Brit., *Trinovantes* (Essex, GB) 51°34′N, 0°48′E
Salt working; briquetage, pottery, terra sigillata, evaporation vessels; 1st to 3rd c.
Nenquin, 1961, 89: *VCH Essex* III (1963), 194–5.

WALESBEECH Brit., *Regni* (East Sussex, GB) 51°05′N, 0°01′W
Iron working; pottery 2nd c.
VCH Sussex III (1935), 31.

WALTON CASTLE Brit., *Trinovantes* (Suffolk, GB) 51°58′N, 1°22′E
?PORTVS ADVRNI or ARDAONI (*qv*).
Fort of the *Litus Saxonicum*, inundated by the sea. *Vicus*, cemetery.
VCH Suffolk I (1911), 305–7: *Arch J* XCVII (1940), 137–40: Cunliffe, 1968, 270: Collingwood-Richmond, 1969, 49: Johnson, 1979, 41–3, 68–71.

WALTON HEATH Brit., ?*Regni* or *Atrebates* (Surrey, GB) 51°16′N, 0°14′W
Villa; hypocausts, tessellated and mosaic pavements. ?2nd c.
VCH Surrey IV (1912), 369: *SyAC* LI (1949), 57–64: Rainey, 1973, 150.

WALTON-ON-THE-HILL Brit., ?*Regni* or *Atrebates* (Surrey, GB) 51°17′N, 0°15′W
Villa, corridor. Baths. Timber house built *c* AD 100 on site of former Iron Age occupation; masonry villa built *c* 180, lasted to *c* 400.
SyAC LI (1949), 65–81.

WALTON-ON-THE-NAZE Brit., *Trinovantes* (Essex, GB) 51°51′N, 1°17′E
Salt working; briquetage, pottery.
CAGQB II (1959), 26–27: Nenquin, 1961, 91: *VCH Essex* III (1963), 198.

WANBOROUGH Brit., *Dobunni* (Wiltshire, GB) 51°34′N, 1°43′W
?DVROCORNOVIVM.
It. A. 485.5.
PNRB 350.
Major settlement, occupied *c* AD 70 to 4th c. Foundations, pavements, iron slag, many small finds, coins including one group of *c* 2000. *Mansio*.
WANHM LXIII (1968), 110; LXV (1970), 204–5; LXVI (1971), 188–9: Rodwell-Rowley, 1975, 233–5: *Britannia* VIII (1977), 223–7; XI (1980), 115–126.

WAREHAM Brit., *Durotriges* (Dorset, GB) 50°41′N, 2°06′W
Minor settlement. Apparently continuous occupation from the Iron Age to modern times. Occupation debris and many small finds.
RCHM Dorset II part 3 (1970), 614.

*****WARFUSÉE-ABANCOURT** Bel., *Ambiani* (Somme, F) 49°52′N, 2°35′E
Large villa.
Agache, fig. 103, 104, 221, 473–75, 636: Agache-Bréart, 1975, 131: Agache, 1978, *passim*.

WATERGATE HANGER Brit., *Regni* (West Sussex, GB) 50°54′N, 0°54′W
Villa, corridor. Date uncertain. Baths, tessellated floors, tiles, pottery.
VCH Sussex III (1935), 28–29.

WAULUD'S BANK Brit., *Catuvellauni* (Bedfordshire, GB) 51°54′N, 0°27′W

Pottery kiln, 2nd c.
VCH Bedfordshire II (1908), 8.

WELLHOUSE Brit., *Atrebates* (Berkshire, GB) 51°27'N, 1°15'W
Villa; small finds including coin of Tetricus (AD 267–74).
VCH Berkshire I (1906), 209.

WELLOW Brit., *Belgae* (Avon, GB) 51°19'N, 2°23'W
Villa, courtyard. Numerous mosaics; coins of 3rd and 4th c.
VCH Somerset I (1906), 312–14: Rainey, 1973, 151–52.

WELWYN Brit., *Catuvellauni* (Hertfordshire, GB) 51°50'N, 0°13'W
Large settlement, probably originating in at least 1st c BC if not earlier and carried on in Romano-British times; coins from Antoninus Pius to Gratian. Foundations and numerous small finds, cemeteries.
VCH Hertfordshire IV (1914), 165–69.

WEMBERHAM *cf* Yatton.

WEST BLATCHINGTON Brit., *Regni* (East Sussex, GB) 50°51'N, 0°11'W
Villa; aisled house. Coins of Tetricus.
VCH Sussex III (1935), 50. *Sx. Arch Colls* LXXXIX (1950), 1–56; Cunliffe, 1973, 84, 106, 112.

WEST CHALLOW Brit., ?*Dobunni* (Oxfordshire, GB) 51°35'N, 1°28'W
Villa, corridor. Hypocausts; coins from Trajan to Constantine.
Arch J XXXIII (1876), 382–92.

WEST COKER Brit., *Durotriges* (Somerset, GB) 50°55'N, 2°41'W
Temple or villa, apparently on site occupied in pre-Roman times. Pottery including samian; coins mid-2nd to 4th c (Lucilla-Valens); statuette of Mars and bronze plaque bearing dedication to Mars Rigisamus.
VCH Somerset I (1906), 331: *RIB* 187: Ross, 1967, 175.

WEST DEAN Brit., *Belgae* (Hampshire, GB) 51°03'N, 1°38'W
Villa, probably courtyard. Baths, mosaics; coins from Victorinus to Magnentius.
VCH Hampshire I (1900), 311–12: *VCH Wiltshire* I (1957), 119: Rainey, 1973, 152–53.

WESTLAND *cf* Yeovil.

WEST MEON Brit., *Belgae* (Hampshire, GB) 51°01'N, 1°06'W
Villa, courtyard. Hypocausts, mosaics.
PHFC V (1904–05), 271: *Arch J* LXIV (1907), 1–14: Rainey, 1973, 153–54.

WEST MERSEA Brit., *Trinovantes* (Essex, GB) 51°46'N, 0°53'E
Mausoleum, diameter 18 m; hexagonal central chamber, six radial walls, 12 buttresses. Globular glass urn with lead cover, lamp; 2nd c. Villa, unknown type; plaster, mosaics, tombs, tumulus.

VCH Essex III (1963), 159, 168–70: Todd, 1978, 19–21.

WESTON UNDER PENYARD Brit., *Dobunni* (Herefordshire, GB) 51°55'N, 2°31'W
ARICONIVM.
It. A. 485.3.
PNRB 257.
Large settlement; extensive occupation during which the early 4th c was the most flourishing period. Iron-working centre. Foundations, iron ore and slag, pottery, many small finds, coins from Claudius to Gratian.
VCH Herefordshire I (1908), 187–90: Jack, G. H., *Excavations on the Site of Ariconium*, Hereford, 1923: *TWNFC* XXXVIII (1964–66), 124–35.

WEYCOCK HILL Brit., *Atrebates* (Berkshire, GB) 51°29'N, 0°49'W
Romano-Celtic temple; octagonal, within temenos. Small finds include coins mostly of 3rd and 4th c.
VCH Berkshire I (1906), 216–18: *Berks AJ* LV (1956–57), 48–68; Lewis, *passim*.

WHATLEY *cf* Nunney

WHEATLEY Brit., *Catuvellauni* (Oxfordshire, GB) 51°44'N, 1°07'W
Villa; baths; coins of late 3rd and early 4th c.
VCH Oxfordshire I (1939), 322–3.

WHERWELL Brit., *Belgae* (Hampshire, GB) 51°10'N, 1°28'W
Villa, corridor. Half-timbered structure; tessellated floors. Probably built *c* AD 300 and destroyed soon afterwards.
JRS LIV (1964), 174; LV (1965), 217.

WHITCHURCH Brit., *Dobunni* (Herefordshire, GB) 51°51'N, 2°40'W
Iron working.
TWNFC XXXVI (1958–60), 227–33.

WHITE NOTLEY Brit., *Trinovantes* (Essex, GB) 51°50'N, 0°35'E
Mausoleum; circular, diameter 2.40 m, 3 or 4 buttresses, central chamber tiled, 1.2 m square. Coins of 2nd c, toilet articles, glass.
VCH Essex III (1963), 164.

WHITE STAUNTON Brit., *Durotriges* (Somerset, GB) 50°54'N, 3°01'W
Villa; baths, hypocausts.
VCH Somerset I (1906), 334.

WHITE WALLS *cf* Easton Grey

WHITMEAD *cf* Tilford

WHITTINGTON COURT Brit., *Dobunni* (Gloucestershire, GB) 51°53'N, 1°59'W
Villa; earliest building a bath-house of early 2nd c; corridor villa probably built early 4th c, mosaics installed *c* 370. Latest coins of Honorius, AD 393–423.
TBGAS LXXI (1953), 13–87: Rainey, 1973, 155–56.

WHITTON Brit., *Silures* (South Glamorgan, GB) 51°26'N, 3°19'W
Farmstead. Pre-Roman circular timber houses

succeeded by further round houses and then in 2nd c. by rectangular buildings with stone foundations; occupation ceased c. AD 340. Numerous small finds include painted plaster, pottery, glass, bronzes.

JRS LVI (1966), 196; LVII (1967), 174; LVIII (1968), 176; LIX (1969), 200. Jarrett, M. G. and Wrathall, S., *An Iron Age and Roman Farmstead in South Glamorgan,* Cardiff, 1981.

WICK Brit., *Belgae* (Avon, GB) 51°27'N, 2°25'W

Villa; hypocausts. Coins of 3rd and 4th c.

PBNHAFC I (1868), 1–16.

WIGGINTON Brit., *Dobunni* (Oxfordshire, GB) 51°59'N, 1°26'W

Villa; hypocausts, mosaics. Coins Victorinus to Valens.

VCH Oxfordshire I (1939), 309: *JRS* LVI (1966), 208: Rainey, 1973, 157.

WIGGONHOLT Brit., *Regni* (Sussex, GB) 50°57'N, 0°29'W

Villa (bath-house) on Lickfold Farm; lead tank inscr. chi-rho found 450 m to NW.

SxAC LXXVIII (1937), 13–36; LXXXI (1940), 55–67; *Ant. J.* XXIII (1943), 155–7; *Britannia* XII (1981), 271–6.

WILCOTE *cf* Shakenoak

WIMEREUX Bel., *Morini* (Pas-de-Calais, F) 50°46'N, 1°37'E

Pottery kilns (in the foundations of the railway station).

Leduque, *Boulonnais,* 95, VIII, n. 63.

WINCHESTER Brit., *Belgae* (Hampshire, GB) 51°03'N, 1°19'W

VENTA BELGARVM.

Ptol. II,3,13 (Οὐέντα).

It. A. 478.2, 483.2 (*Venta Belgarum*), 486.11 (*Venta Velgarum*).

?*Not. Dig. Occ.* XI, 60 (*procurator gynaecii . . . Ventensis*).

Rav. Cos. 106.18 (*Venta Velgarom*).

Beda *Ecc Hist* III,7; IV, 15; V,18; V,23.

PNRB 492.

Town, capital of the CIVITAS BELGARVM, partly overlying Iron Age defended settlement. Claudian timber buildings; earthen rampart (?late Neronian or early Flavian) enclosing c 58 ha, replaced by stone wall c 200. Forum and basilica, Romano-Celtic temple, shops and houses. The imperial wool-mill (*gynaecium*) mentioned by *Not. Dig.* was probably at W., though sometimes claimed for VENTA ICENORVM (Caistor St Edmund, Norfolk). Outside the town, buildings and cemeteries, including important 4th c one at Lankhills.

Cunliffe, B. (ed), *Winchester Excavations, 1949–60,* W., 1964: Biddle, M., in *Ant J* XLIV (1964) – LV (1975): *Latomus* XXVI (1967), 648–76: Rainey, 1973, 157–8: Wacher, 1975, 277–88: Clarke, G. (ed), *The Roman Cemetery at Lankhills,* Oxford, 1979. *Britannia* I (1970), 301; II (1971), 283–4; III (1972), 348–9; IV (1973), 318–20; VI (1975), 278–9; VII (1976), 371–2; VIII (1977), 419; IX (1978), 465–6; X (1979), 331–2; XI (1980), 395–6.

WINGHAM Brit., *Cantiaci* (Kent, GB) 51°16'N, 1°13'E

Villa, unknown type; bath, painted plaster, hypocausts, mosaics, mural mosaics.

Ant XVIII (1944), 52–55: *Arch Cant* LXXXII (1967), lx: Rivet, 1969, 153: *VCH Kent* III (1932), 125.

WINGLES Bel., *Atrebates* (Pas-de-Calais, F) 50°29'N, 2°51'E

Gallo-Roman foundations.

Jelski, 142.

WISLEY Brit., ?*Regni* or *Atrebates* (Surrey, GB) 51°19'N, 0°29'W

Pottery kiln; 1st c.

SyAC XXV (1912), 131–32.

WISSANT Bel., *Morini* (Pas-de-Calais, F) 50°53'N, 1°40'E

?PORTVS ITIVS.

Caes. *BG* V, 2; V, 5 (ed. Rambaud, 1974).

Str. IV,5,2 (C.199) (τὸ Ἴτιον).

Ptol. II,9,1 (Ἴτιον ἄκρον).

CIL XIII.1.2, p 561.

Port, probably the *Portus Itius* used by Caesar in his invasions of Britain 55–4 BC. Iron Age oppidum ('Camp de César', 'Motte Juliette', 'Mont du Châtel') occupied under the early Empire; cemetery, coins, pottery.

Haigneré, D., *Étude sur le Portus Itius de Jules César,* Paris, 1862: *Bull Soc géo Lille,* March 1893: *BSAB* VIII (1908–9), 462–7: P-W IX.2 (1916), 2367–70; XXII.1 (1953), 407: Briquet, A., *Le littoral du N de la France et son évolution morphologique,* Paris, 1930: Grenier, *Manuel* II.2 (1934), 512, 514, 527–9: *CRAI,* 1944, 372–86: Leduque, *Boulonnais,* 33; *Morinie,* 71: *RÉL* XLI (1963), 186–209: *Celticum* XV (1965), 53–95: Ringot, 168: *Septentrion* II (1972), 27–31: Delmaire, 1976, 94–8; *Procs. Brit. Academy* LXIII (1977), 150–2.

WISSOUS Lug., *Parisii* (Essonne, F) 48°44'N, 2°20'E

Aqueduct of Arceuil.

Duval, 1961, 173–6: Roblin, 1971, 181, 287.

WISTON Brit., *Regni* (West Sussex, GB) 50°54'N, 0°21'W

Tile kiln.

SxAC II (1849), 313–15: Cunliffe, 1973, 118–9.

WITHINGTON Brit., *Dobunni* (Gloucestershire, GB) 51°50'N, 1°57'W

Villa, probably courtyard. Many mosaics.

Arch XVIII (1817), 118–121: Finberg, H. P. R., *Roman and Saxon Withington,* Leicester, 1955: Rainey, 1973, 162. RCHM, *Gloucestershire Cotswolds,* 1976, 131–2.

WITHYHAM Brit., *Regni/Cantiaci*? (East Sussex, GB) 51°05'N, 0°08'E

Iron workings; wooden buildings, six furnaces, pottery 2nd to 4th c.

KAR IV (1966), 12–14: Straker, 1931, 257.

WIVELISCOMBE Brit., *Dumnonii* (Somerset, GB) 51°02′N, 3°18′W
Fort, probably Claudian; 1st-c pottery.
PSANHS CIII (1958–59), 81–89.

WOODCHESTER Brit., *Dobunni* (Gloucestershire, GB) 51°44′N, 2°14′W
Large villa; two courtyards. Many mosaics including the magnificent 'Great Pavement' which has been on show at long intervals. Pottery and coins of 2nd, 3rd and 4th c.
Lysons, S., *An Account of Roman Antiquities discovered at Woodchester*, London, 1797: *TBGAS* XLVIII (1927), 75–96; LXXIV (1956), 172–5: Smith, D. J., *The Great Pavement and Roman Villa at Woodchester*, Gloucester, 1973: Rainey, 1973, 163–4: *Britannia* V (1974), 451; XIII (1982), 197–288: McWhirr, 1981, 94–5.

WOODEATON Brit., *Catuvellauni* (Oxfordshire, GB) 51°48′N, 1°13′W
Romano-Celtic temple in temenos on pre-Roman site. Rectangular foundations; numerous votive objects including bronze birds, letters, figures, etc. In continuous use from Flavian times to the early 5th c.
Oxon XIV (1949), 1–45; XIX (1954), 15–37: Lewis, 1966, *passim*: Ross, 1966, *passim*.

WOODSPEEN *cf Spinis*

WOOLASTON Brit., *Dobunni* (Gloucestershire, GB) 51°41′N, 2°35′W
Villa; baths, hypocausts. Occupied *c* AD 130 to 4th c.
Arch Camb XCIII (1938), 93–125; CII (1953), 100: McWhirr, 1981, 97–9.

WOOLSTONE Brit., *Dobunni* (Oxfordshire, GB) 51°35′N, 1°35′W
Villa, probably corridor. Tiles, tesserae, mosaics, pottery of 2nd to 4th c.
VCH Berkshire I (1906), 222: *Berks AJ* LVII (1959), 83–85: Rainey, 1973, 165.

WORBARROW BAY Brit., *Durotriges* (Dorset, GB) 50°37′N, 2°11′W
Shale working.
RCHM *Dorset* II part 3 (1970), 612.

WORPLESDON Brit., ?*Regni* or *Atrebates* (Surrey, GB) 51°15′N, 0°37′W
Villa; mosaic.
VCH Surrey IV (1912), 370: Rainey, 1973, 165.

WORTH Brit., *Cantiaci* (Kent, GB) 51°14′N, 1°21′E
Temple in its temenos on the site of an Iron-Age temple.
Ant J XX (1940), 115–21: Lewis, 1966, 3: Ross, 1967, 46, 198.

WORTHING Brit., *Regni* (West Sussex, GB) 50°49′N, 0°23′W
RIB 2220.
Milestone of Constantine I (AD 307–37).
Sedgley, 1975, 18.

WORTH MATRAVERS Brit., *Durotriges* (Dorset, GB) 50°36′N, 2°02′W
Shale working. Coin of Victorinus (AD 270).
RCHM *Dorset* II part 3 (1970), 621.

WRAXALL Brit., *Belgae* (Avon, GB) 51°26′N, 2°45′W
Villa; occupied *c* AD 250–350. Large bath suite.
PSANHS CV (1961), 37–51.

WYCOMB Brit., *Dobunni* (Gloucestershire, GB) 51°53′N, 1°58′W
Minor settlement, certainly occupied late 3rd to early 5th c, possibly earlier. Foundations including those of two successive temples, ?three other ?shrines, and ?theatre. Earlier temple 9 m×11 m divided internally into two; later temple square 13 m, raised cella floor. Columns, fragments of sculptures, reliefs, bronze 'Mars', votive objects, pottery, coins.
TBGAS LXXVIII (1959), 161: *Britannia* IV (1973), 311: Branigan-Fowler, 1976, 112–4: McWhirr, 1981, 70–73. RCHM, *Gloucestershire Cotswolds*, 1976, 125–6.

WY-DIT-JOLI-VILLAGE Lug., *Veliocasses* (Val-d'Oise, F) 49°06′N, 1°50′E
Lime kiln.
Gallia XVII (1959), 276; XIX (1961), 291.

WYKEHURST Brit., *Regni* (Surrey, GB) 51°09′N, 0°27′W
Tile kiln, 1st and 2nd c.
SyAC XLV (1937), 74–96: Cunliffe, 1973, 119.

YATTON Brit., *Belgae* (Avon, GB) 51°23′N, 2°48′W
Romano-Celtic temple. 1, rectangular building of uncertain date; 2, succeeded by another rectangular building coeval with an octagonal building *c* 9 m diameter of late 2nd–early 3rd c; 3, a square Romano-Celtic temple (*c* 15 m) of first half 3rd to end 4th c. All within a square temenos. Pottery 1st to 4th c, coins 250 to 400, bronze figurine.
JRS LV (1965), 216; LIX (1969), 227: *Britannia* I (1970), 295.

YATTON Brit., *Belgae* (Avon, GB) 51°23′N, 2°51′W
Villa; baths, mosaics. Occupied 3rd and 4th c.
VCH Somerset I (1906), 306: *JRS* LV (1965), 216: Rainey, 1973, 167.

YEOVIL Brit., *Durotriges* (Somerset, GB) 50°56′N, 2°39′W
Villa; courtyard. Baths, many mosaics. Occupied late 2nd to late 4th c.
PSANHS LXXIV (1928), 122–43: Rainey, 1973, 153.

YÈVRE-LE-CHÂTEL Lug., *Carnutes* (Loiret, F) 48°09′N, 2°20′E
Gallo-Roman foundations, querns.
Nouel, 30, 31.

YEWTREE FARM *cf* Congresbury

YFFINIAC Lug., *Coriosolites* (Côtes du Nord, F) 48°29′N, 2°41′W

'la Ville-Volette', 'le Val', 'Champ de la Bail'; Gallo-Roman foundations, coin hoards. Tegulae, potsherds, burials, traces of fire. Pot containing hundreds of small bronze coins of the 3rd c; two pots containing 800 small bronzes for the most part Constans, Constantine I, Valentinian (museum of Saint-Brieuc); coins of 1st to 3rd c.
BSCN V (1867), 51, appendix: *MSCN* I (1883–84), 174: *ACN,* 1939.

Y GAER (=Brecon Gaer) Brit., *Silures* (Powys, GB) 51°57′N, 3°27′W
?CICVCIVM
Rav. Cos. 106.26 (*Cicutio*).
RIB 403–5, 2258–9.
PNRB 307.
Fort, occupied *c* AD 75–200, together with small settlement.
Nash-Williams, 1969, 48–51.

YMARE Lug., *Veliocasses* (Seine-Maritime, F) 49°21′N, 1°11′E
Villa; debris of columns, mouldings, pottery.
BSNÉP XVII (1909), 32–38.

YMONVILLE Lug., *Carnutes* (Eure-et-Loir, F) 48°15′N, 1°45′E
Gallo-Roman foundations, mosaic, terra sigillata, coins. 'Mérouvilliers', coin hoard. 'Rosay', foundations, terra sigillata.
Nouel, 16, 36.

YOUNGSBURY Brit., *Catuvellauni* (Hertfordshire, GB) 51°50′N, 0°01′W
Two barrows *c* 18 m diameter 3.5 m high. One excavated contained wooden cist with hinged lids *c* 1 m × 0.5 m; pottery jug, urn containing cremated bones and 200 nails, glass vessel with cremated bones. Late 2nd c.
VCH Hertfordshire IV (1914), 164.

Y PIGWN Brit., *Silures* (Powys, GB) 51°58′N, 3°42°W
Two temporary camps.
JRS XLIII (1953), 86: Nash-Williams, 1969, 124–125.

YSTRADFELLTE Brit., *Silures* (Powys, GB) 51°50′N, 3°34′W
Temporary camp.
BBCS XXI (1964–66), 174–78: Nash-Williams, 1969, 125–126.

YVIGNAC Lug., *Coriosolites* (Côtes du Nord, F) 48°21′N, 2°11′W
'la Boissière', Gallo-Roman foundations exposed in 1864; tegulae, painted plaster, coin of Carinus, 3rd c.
MSCN, 1874, vi.

ZOTEUX Bel., *Morini* (Pas-de-Calais, F) 50°37′N, 1°53′E
Cemetery, near the church; altar to Jupiter, coins.
Leduque, *Boulonnais*, 113: Delmaire, 1978, 341.

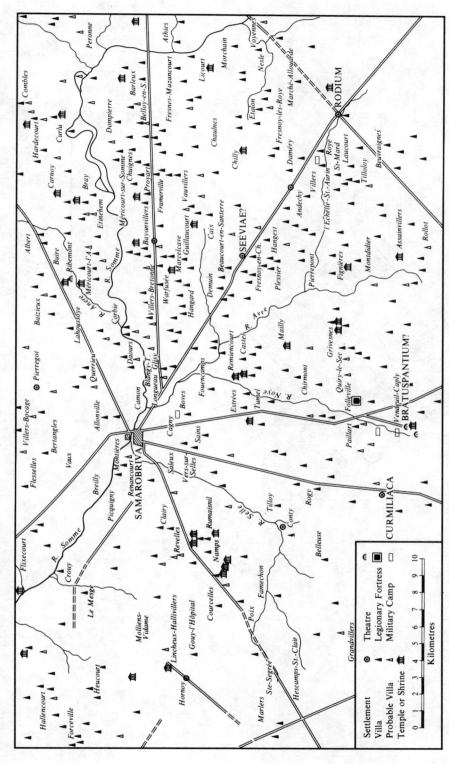

Map showing the distribution of villas and shrines around Amiens. For full particulars of this and surrounding areas see R. Agache and B. Bréart: *Atlas d'archéologie aérienne de Picardie: la Somme protohistorique et romain*, Amiens, 1975 and R. Agache: *La Somme pré-romaine et romaine*, Amiens, 1978.

CAERLEON

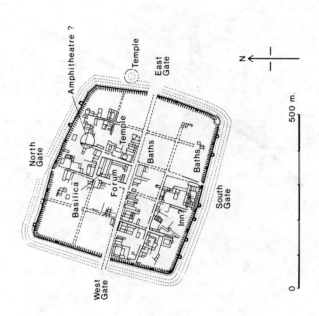

CAERWENT

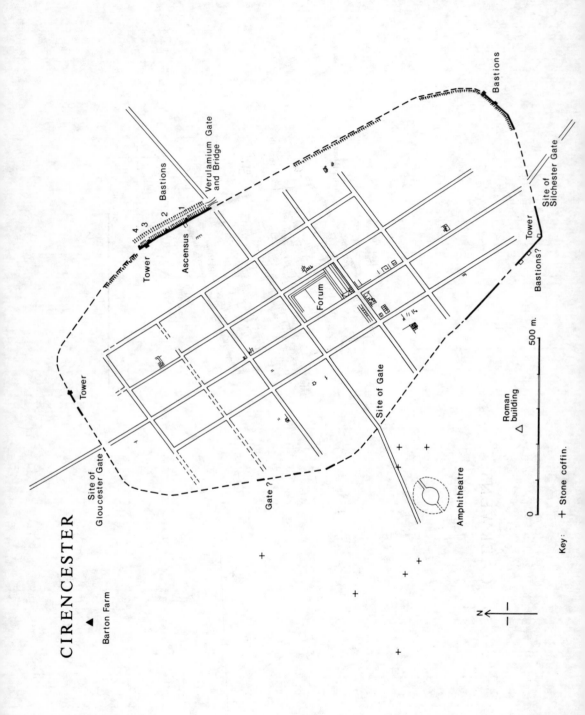

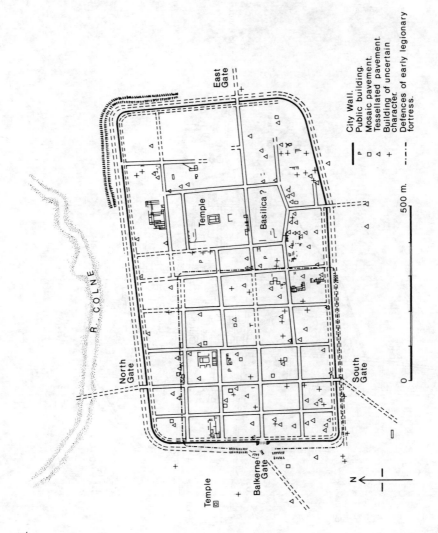

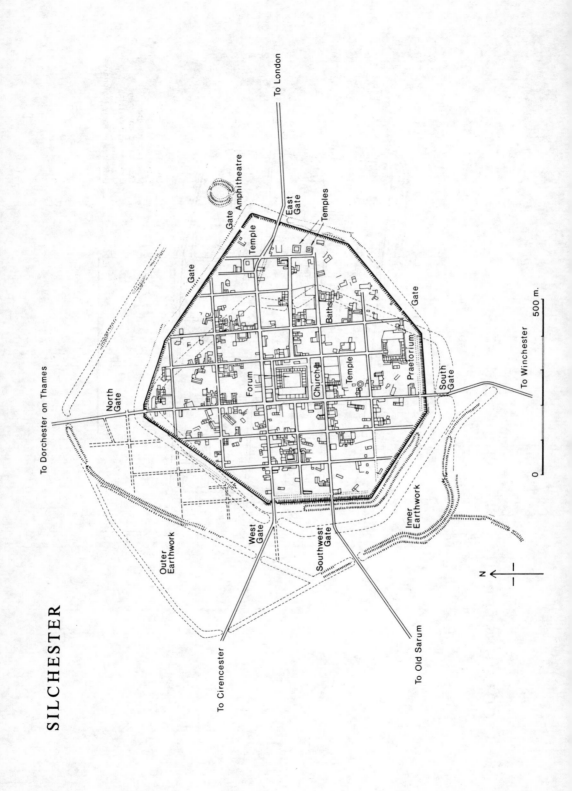

VERULAMIUM

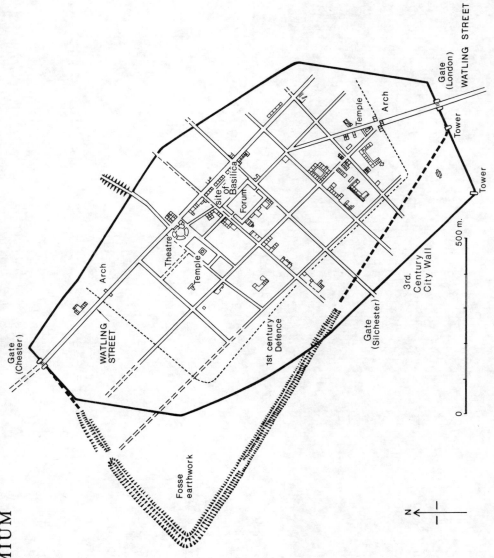

MEMORANDUM